T0296675

ALL SCIENTISTS NOW

The Deputation to Faraday (May 1857).
From left to right: Lord Wrottesley (P.R.S.);
J. P. Gassiot (1797–1877), F.R.S. 1840; Sir William Grove (1811–96),
F.R.S. 1840; Michael Faraday (1791–1867), F.R.S. 1824.
By E. Armitage

ALL SCIENTISTS NOW

The Royal Society in the nineteenth century

MARIE BOAS HALL

Emeritus Reader in History of Science and Technology
Imperial College, London University

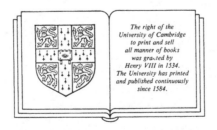

The right of the
University of Cambridge
to print and sell
all manner of books
was granted by
Henry VIII in 1534.
The University has printed
and published continuously
since 1584.

CAMBRIDGE UNIVERSITY PRESS

Cambridge

London New York New Rochelle

Melbourne Sydney

Published by the Press Syndicate of the University of Cambridge
The Pitt Building, Trumpington Street, Cambridge CB2 1RP
32 East 57th Street, New York, NY 10022, USA
296 Beaconsfield Parade, Middle Park, Melbourne 3206, Australia

© Cambridge University Press 1984

First published 1984

Printed in Great Britain by the University Press, Cambridge

Library of Congress catalogue card number: 84-7705

British Library cataloguing in publication data

Hall, Marie Boas
All scientists now.
1. Royal Society – History
I. Title
506'.041   Q41.L85

ISBN 0 521 26746 3

CONTENTS

ILLUSTRATIONS

Reproduced by kind permission of the President and Council of the Royal Society from negatives in their possession

FOREWORD

Dr Marie Hall has done a great service in writing an account of the Royal Society in the nineteenth century. This is a phase of our history which is nearly as dramatic as our much better-known beginnings in the seventeenth century; the events leading up to the contested election of a President in 1830 make the present holder of that office grateful that present-day Fellows do not express their feelings with the freedom and rancour that characterize the public outpourings of certain disgruntled Fellows of a century and a half ago.

The Society entered the nineteenth century as a monarchy in the literal sense: it was ruled by one man, the redoubtable Sir Joseph Banks, then just half way through his forty-two-year tenure of the Presidency, and the Fellowship was divided between active scientists and wealthy amateurs. By 1850, the Society had converted itself to essentially its present-day structure, in which the President has no autocratic power and Fellows are elected solely for scientific distinction. Dr Hall traces this transformation with great skill, and shows that it was a less straightforward, and indeed more interesting, process than one might suppose: the reforming party was defeated in 1830, but reform crept forward and by 1847 the mood was such that the new Statutes were accepted without dissension.

Fellows of the Royal Society will be deeply grateful to Dr Hall for undertaking this task with such success, as will also be a wide range of readers interested either in the development of scientific institutions or simply in a piece of history with much human interest.

April 1984   Andrew Huxley
                                                                 President, R.S.

PREFACE

On 28 November 1660 there was held an informal meeting at Gresham College in the City of London at which those who had then just attended a lecture by the professor of astronomy, Christopher Wren, discussed the formation of an organisation on the pattern of Continental academies 'for the promoting of experimentall philosophy', a design which, when he was told of it, met with the approval of Charles II. Regular weekly meetings began immediately; the members agreed to pay a weekly subscription; and a constitution was drawn up for what, at that point, was called simply 'the Society'. On 15 July 1662 a formal Charter of Incorporation was enacted for 'the Royal Society', while in April 1663 a second charter denominated it 'Regalis Societas Londini pro Scientia naturali promovenda', the Royal Society of London 'for improving naturall Knowledge.* It is thus the oldest continuous *scientific* society in the world still operating under its original charter, and its principal publication, the *Philosophical Transactions*, is the oldest continuous scientific journal.

Its charters gave it status but it was a private society, financed by the contributions of its members, the Fellows, who chose their own officers and Council and elected new Fellows by vote. It has remained private and self-governing ever since, although in the twentieth century it receives some financial assistance from the Government. But, uniquely in modern times, it is both a private body and a national one. Other countries have acquired over the centuries national academies to which it is an honour to belong, but election to which implies obligations to the Government which supports the organisation. The Royal Society alone is without direct government ties, while assisting and advising government departments

---

* So it is given in English in the President's oath in the second charter.
  Contemporaries used indifferently the terms 'improving' and 'promoting'. Since *promoveo* in modern English is usually taken to mean 'promoting', this form has been used since the late nineteenth century, as in *The Record of the Royal Society of London*, although until then, and recently, the word 'improving' was used.

when requested to do so; unusually its Fellows pay for the privilege of their membership, rather than being paid by their Government. Nowadays the Royal Society acts as a national academy without being one, a position it attained during the nineteenth century. (National academies nearly all comprise a number of bodies devoted to various disciplines, of which natural science is only a part; the Royal Society, in contrast, confines its interest to what for nearly 150 years has been called in English 'science', but which other languages denominate 'natural science'. True, in 1660 'natural knowledge' was loosely defined, while during the two subsequent centuries many Fellows were primarily distinguished for their interest in the disciplines we now call archaeological, numismatic, historical, anthropological or antiquarian, or their interest in natural science was as patrons or dilettanti rather than as practitioners. Only in the course of the nineteenth century did the Society determine to restrict its membership to those who actively practised science as it was coming to be defined in English and had, moreover, achieved some eminence in it.) This all rather increased the importance of the Society in the intellectual life of Great Britain than otherwise, and caused it to play a more influential rôle in regard to the Government than had been the case in earlier times.

An examination of the ways in which the Royal Society effected these changes and the influence which these changes had on the Society itself is thus of some general historical interest. It is also of interest for anyone seeking to understand that curious and uniquely English political practice, the art of changing while remaining the same. For the Royal Society transformed its whole administrative structure without any change in the charters which govern it, under which it exists as happily now as it did in 1663. Moreover, it emerged from a period of drastic change in the mid-nineteenth century with an enhanced reputation, the acknowledged chief of all British scientific societies, its President eminent by reason of his office, its position in the world of science unchallengeably supreme. It was a typically Victorian achievement in a period of British hegemony and self-confidence.

The Royal Society's early history and even its pre-history has been discussed at length by many historians in recent times. Its eighteenth-century history has been less studied, mainly because of an English conviction that English science did not thrive in this period; pending a modern re-appraisal one can gain a fair idea of the Society's 'life and work' from the detailed account in *A History of the Royal Society* (1848) by C. R. Weld, then Assistant Secretary, who made extensive use for the first time of original documents. Weld, modestly and realistically, stopped his work in 1830 and there has been little study of the post-1830 period since, with the exception of a few articles and two major works. *The Record of the Royal Society of London* (first edition 1897, fourth 1940) contains much extremely useful historical information, including lists of Fellows. The fourth edition was prepared by Sir Henry Lyons, Treasurer 1929–39; it was presumably the stimulant for his 1944 *The Royal Society, 1660–1940*, intended to be, as its sub-title calls it, 'A History of its Administration under its Charters'. This valuable work is closely administrative and financial, full of detail but lacking historical subtlety, too dense for casual reading. The present work, the utility of which was first suggested to me by N. H. Robinson, has tried to strike a balance between precision and interest. It is intended to be readable by Fellows interested in the past of the Society of which they form a part as it developed during the nineteenth century.

I have dealt in some detail with the administrative problems of the early part of the nineteenth century because they were of general public interest, and because controversy is historically interesting to examine and of absorbing interest to most people. I have dwelt in less detail on such history in the latter half of the century, partly because things went smoothly, which is less interesting than when they do not, and partly because the problems were so like modern ones that they do not need to be so carefully explained. For that period I have preferred to discuss the Royal Society's influence upon science, its relations with Government and with other societies and its interest in promoting exploration.

If I have written anything which scientists find inexplicable or abnormal, I apologise; I am a historian, not a scientist nor (conceivably) a Fellow. To historians I must apologise because this is traditional history. I have deliberately avoided either statistical, prosopographical studies or socio-logical studies as unsuitable for my purposes; but I should be pleased if others find here some material which will be either useful or stimulating in such studies.

This is in no sense an official history. I have to thank the President and Council of the Royal Society for permission to use their archives freely and to reproduce such portions of them as I chose, as also for the generosity with which they gave me funds to cover major expenses. I have been given access to all the documents which I asked to see and had others pointed out to me; for this I thank warmly the Librarian and his staff who patiently fetched and carried on my behalf, especially the former archivist, Mr Leslie Townsend, who from his unrivalled knowledge of the Library's resources made my first steps so much easier than they would otherwise have been. A number of errors in detail arising from my inevitable ignorance as an outsider have been kindly corrected by the Executive Secretary, Dr R. W. J. Keay, C.B.E. I particularly wish to thank Professor Sir William Paton, F.R.S., who read the whole manuscript sympathetically and made valuable constructive suggestions. That there will inevitably be errors left is my fault alone.

I am grateful to Commander Derek Howse and to Dr Stuart Malin who answered questions about navigation and magnetism, respectively, and to the Libraries of the British Museum, the Wellcome Institute and University College London. I am most grateful to Mrs Julie Hounslow who patiently typed my sometimes very rough drafts. And above all I thank my husband, Rupert Hall, who has helped in numerous ways, not least by great forbearance when I was preoccupied with another century's actions.

Tackley, Oxford   Marie Boas Hall
October 1983

Fig. 1. Sir Joseph Banks, Bart (1744–1820), F.R.S. 1766, P.R.S.
1778–1820.
By Thomas Phillips, 1815

1

The eighteenth-century legacy

When the nineteenth century opened, there was no reason to suppose that the Royal Society would change from its eighteenth-century pattern, nor indeed was there any reason to suppose that it needed change. Sir Joseph Banks, rich, influential and eminent, was almost exactly at the mid-point of his forty-two-year term as President. This fact, as well as his rôle as patron of learning, both intellectually and socially, strikingly illustrates the leisurely atmosphere of intellectual life that existed in the Royal Society as in the literary and philosophical societies of the provinces in an age before intellectual specialisation and the social development of Club life. Indeed, the Royal Society, founded 'for improving naturall Knowledge' in the mid-seventeenth century, had had little reason or impetus to change the views of its founders, who combined a Baconian belief that any willing, literate man might contribute to that improvement with the practical understanding that a society dependent for its finances upon the contributions of its members needed a reasonable proportion of the rich and the influential to survive.

The election as President in 1778 of Joseph Banks (F.R.S. 1766) – young, wealthy, with the glamour of his voyage with Cook still alive, botanist and generous patron of botanists, a friend of George III – perfectly conformed to the seventeenth-century as to the eighteenth-century ideal, and as a result the Society could look forward with some certainty to a period of prosperity. Crown favour led to the offer of relatively spacious apartments in Somerset House, to which the Society moved in 1780 (from the now cramped quarters in Crane Court, purchased in Newton's Presidency). Banks's own position as patron of science flourished too, and every country and foreign visitor with any intellectual pretensions was flattered to be able to attend Banks's Sunday evening receptions and his lavish and stimulating breakfast parties, all of which might lead to election as F.R.S., either domestic or foreign.

That Banks was made a baronet in 1781 gave even more lustre to his patronage, and to the Society over which he presided with a firm hand.

His dominance was challenged only once, in 1784, and it is difficult to analyse the exact position of the rebels.[1] The protagonist was Charles Hutton, who taught mathematics at Woolwich; he had been elected Foreign Secretary of the Society in 1779, but was dismissed by Banks ostensibly for non-residence in London and, so Banks declared, consequent inability to discharge his office. His chief supporter was Samuel Horsley, then editing Newton's works, a man of minor mathematical reputation, later a bishop. Was this, as some have seen it, a revolt of 'professionals' against 'gentlemen'? Was it rather, as Horsley tried to make it, a revolt of the mathematical scientists against (to use a neologism) the biological scientists, A against B? Certainly under Banks the biological sciences were given greater importance than they were to receive for much of the nineteenth century: Banks began that close association of the Royal Society with the British Museum (then still subsuming natural history and antiquities under one roof) which continued well into the nineteenth century (many of the Society's Secretaries being officers of both institutions, and the President *ex officio* a trustee); he also much assisted James Edward Smith in the founding of the Linnean Society in 1788. Banks triumphantly won control in 1784, and the biological sciences dominated for the rest of the century, in contrast to the later nineteenth-century power struggle which was to be all the other way.

In 1800 Banks, then fifty-seven years old, and in the twenty-second year of his Presidency, could look back with some satisfaction over his 'reign' (as younger contemporaries felt it to be) and forward with confidence. The Society was, in his eyes, in a very flourishing shape, and he was in a position to know: it has been calculated that of 450 Council meetings during his forty-two years as President he attended no fewer than 417,[2] and he insisted on controlling the membership of the Society, scrutinising the proposals for all candidates. And as if to prove the virtues of his rule he was, in 1801, made a member of the newly formed and highly prestigious Institut de France, an honour he much appreciated, although, ironically in the light of the unstinted approval given to the Institut by English scientists a generation later, there was some decided criticism of Banks's acceptance on the grounds that it was unpatriotic to accept honours from an enemy.

What, it well may be asked, did the Society do during the later years of Banks's Presidency, when he was so firmly in control? The Journal Books (containing minutes of the Society's meetings) reveal a pattern for meetings established in the later eighteenth century and not to be disturbed during Banks's lifetime whether he was presiding (as he normally did) or was absent (as occasionally happened in the earliest years of the new century, but hardly otherwise). Since 1780 the meetings had begun at 8 o'clock on Thursday evenings. The minutes record that the strangers present (not more than two might be introduced by each Fellow) were first named (and presumably welcomed), after which the minutes of the previous meeting were read, followed by a list of 'presents' (books, specimens, occasionally instruments and even portraits). Then came the reading of the certificates of proposed new members, before their suspension in the meeting room, and, when appropriate, ballotting for those proposed earlier (only foreign members, limited in number, were elected annually), and the admission of any Fellows previously elected within the past four weeks (the statutes requiring that any Fellow not so presenting himself forfeited his election, though custom permitted petitioning the Council for delay). After this business was completed, a paper was read in full, in abstract or in part, to be continued at later meetings if too long to fit into the time allotted (about one hour).

Thus in his diary under 11 April 1811 Lord Webb Seymour (F.R.S. 1802, younger brother of the 11th Duke of Somerset, F.R.S. 1797, both keen amateurs of science) recorded that he attended 'a meeting of the Royal Society, to hear Mr Playfair read a part of his paper on the Huttonian theory of volcanoes'.[3] Similarly on 26 January 1814 Charles Babbage (F.R.S. 1816) informed his intimate friend John Herschel (F.R.S. 1813) that 'I was last thursday at a Meeting of the Royal Society where a paper of Davy's was read of a highly curious nature';[4] as Babbage was not yet a Fellow he necessarily had been introduced by a Fellow. Papers were sometimes read by the author (if he were a Fellow), usually by one of the Secretaries; when the papers were by Fellows they were sent or handed to the Secretary with or without the President's approval; when not by Fellows they were either submitted by them to the Secretary or 'communicated' directly by the Fellow concerned. There was at this time no discussion of papers, although occasionally there was a movement to have the custom re-introduced; it had been usual until the mid-eighteenth century but was now claimed to be contrary to existing statutes (which were really intended to preclude unplanned discussion of administrative or political matters) and many Fellows thought it undesirable because undignified and possibly introducing heat into what was intended to be a solemn occasion. But the established procedure could be dull. According to John Barrow (F.R.S. 1806, when he had just begun his long career as second Secretary to the Admiralty, junior to J. W. Croker, F.R.S. 1810) 'The subjects were generally scientific [!] and I confess I often found them dull enough in themselves, and not always improved by the monotony of an official reader, sometimes weary of his task'.[5] Papers so read were then considered for publication in the *Philosophical Transactions* by the Committee of Papers, and most (but not all) were published.

Barrow clearly preferred the social side of membership in the Royal Society: breakfast or 'conversation' with the President, tea-table conversation after the meetings, and dining with other members of the Royal Society Club before the meeting, at 6 o'clock. This Club was limited in membership (forty Fellows by 1820), strangers at first being invited only by the President, although later by other members. Elected members who were not regular in attendance were dropped. Its social composition was varied, by no means confined to Court or Government circles – Marc Isambard Brunel (F.R.S. 1814) was a member in 1816, as Lord Webb Seymour informed his brother.[6] Barrow, an enthusiastic but totally nonscientific Fellow (although he had travelled widely, his appreciation of the geography of regions unknown to him was decidedly limited) all unconsciously illustrated one of the weaknesses of the Royal Society of his time, namely that its composition was far from uniformly devoted to any specific pursuit. For although all Fellows were supposed to be *interested* in the world of learning, it was by no means thought requisite that they be *practitioners* of learning. The seventeenth-century term 'natural knowledge' covered a wider range than the word 'science' was to assume after 1830 or thereabouts, and in this period the English-speaking world was still close to Continental usage, the Royal Society including far more than pure science within its range, just like the French Institut, or various German 'Wissenschaftlichen' societies and academies. The qualification of many a would-be Fellow of the first decade of the nineteenth century was that he was a member of the Society of Antiquaries or possessed wide literary knowledge; naturally the certificates of these candidates were signed by those Fellows of similar qualifications, while the more scientific members signed the certificates, more usually, of scientific candidates.[7] Banks himself took a wide view, although he certainly preferred a candidate to be interested in natural science. As Benjamin Brodie (F.R.S. 1810, and a most eminent surgeon) later recalled, 'The view which Sir Joseph Banks took of the construction of the Society was, that it shall consist of two classes: – the working men of science, and those who, from their position in society or fortune, it might be desirable to retain as patrons of science'.[8] And he instanced the occasion when Sir Everard Home (F.R.S. 1787) proposed as candidate a fashionable physician whom Banks refused to accept, although after the man inherited a fortune and a title Banks declared him to be acceptable as in a position to act as a patron of science. There were also some other groups whose members, like Croker and Barrow, were acceptable to Banks because they were men in public life, and involved with affairs which might concern the Society.

Ordinary Fellows of the Society were hardly aware of any close connection with the world of affairs, for the regular meetings were intended for the reading of papers. Nor was this kind of business discussed at the Anniversary Meetings on St Andrew's Day (30 November), which were primarily devoted to the Society's private affairs. The President began these by analysing the state of the Society: Fellows deceased and Fellows elected, new Council members and any changes in the officers. He then announced and explained the award of medals: the Copley Medal, normally awarded annually, and, after 1800, the Rumford Medal, technically awarded every two years (in fact only thirteen times before 1846). The recipients between 1800 and 1820 are a very worthy collection of men of achievement.

The rise of new societies

One of the few deviations from the norm came in 1809, when Banks announced approvingly the formation of a Society for the Promotion of Animal Chemistry which had been conceived as having special links with the Royal Society, so much so that it was to be regarded as an 'associated' society, and all discourses read before it were to be offered for publication in the *Philosophical Transactions*. The minutes of the Committee of Papers show that a fair number of papers were offered by members in the period, mostly on the chemistry of animal secretions, and virtually all were published. Ironically, after this promising start the Society became little more than a dining club and soon died away. This is in direct contrast to the Geological Society which, begun in 1807 at the instigation of William Babington (physician and mineralogist, F.R.S. 1805) to raise a subscription to publish a French work on crystallography, continued as, essentially, a club, primarily interested in mineralogy.[9] Then, two years later, the Hon. C. F. Greville (F.R.S. 1772) proposed that the Geological Society should become an associate of the Royal, a proposal which Banks naturally warmly welcomed, so much as to consider joining the Society. A formal plan for consolidating the Geological with the Royal Society, as an Assistant Society, was drawn up for consideration, very like what was being done for the Society for the Promotion of Animal Chemistry (probably under Banks's guidance). But the members of the Geological Society rejected this outright, resolving

> that any proposition tending to render this Society dependent upon or subservient to any other Society, does not correspond with the

conception this Meeting entertains of the original principles upon which the Geological Society was founded.

(No doubt the notion of two classes of members, those who were and were not F.R.S., had a good deal to do with the rejection but the question of publication was also a thorny one.) Banks, receiving a copy of this and other resolutions, was indignant, but clearly the geologists were right to maintain their independence: they continued to have a thriving although not totally scientific or professional society, many of whose members were also Fellows of the Royal Society. But it must be added that a number of prominent members of the Geological Society, notably Roderick Murchison (F.R.S. 1826) and Leonard Horner (F.R.S. 1813), were to be active in attempts to revolutionise the Royal Society later on: had he lived to see it Banks would have felt justified in his prognostications of the evil effects of the creation of specialised societies. He had of course not objected to the formation of the Linnean Society, but that had been organised by a friend, and he himself had been active in it from the beginning, for it dealt with his favourite science, botany. His attitude to the Royal Institution seems to have varied: he had taken part in its foundation, and in 1820 John Herschel could declare[10] that it was said that 'the R.I. could at all times command a Secretaryship of the Royal Society', but he seems at times to have shown some jealousy of its independence.

Banks's disapproval was shown clearly and positively in 1820, the last year of his life, when the Astronomical Society was formed.[11] Then he actively interfered in its affairs, dissuading both the Duke of Somerset and Davies Gilbert from accepting the presidency. He apparently told Somerset that the new 'Society will be *the ruin* of the Royal Society', causing Francis Baily (F.R.S. 1821 and later Treasurer), one of the founders of the Astronomical Society, to remark drily that this was 'no mean compliment to us, but not very respectful to that learned body – '. As Baily protested to Babbage,

> It surely cannot be maintained for a moment, that, because a person is a member of the Royal Society, he is precluded from joining any other Society which has Science for its object: & after the fruitless and *more violent* attempt, which Sir Joseph made against the Geological Society, & the Royal Institution (and which only tended to unite more firmly the original members), I wonder that he should again endeavour to oppose the progress of science in this particular instance.

Banks had already tried to influence Davies Gilbert, who liked the idea of being president of the new society but frankly told Babbage

In the course of my short visit to London I have had an opportunity of ascertaining the opinion of several leading Members of the Royal Society, and I have [learned] not without regret that some degree of Jealousy is clearly to be entertained of our new establishment, so that it would be thought, by many Persons, a dereliction of duty, if their Vice President should take a leading part elsewhere.

But he firmly remained a member. It is true that among the founders of the Astronomical Society were many who disapproved of Banks and all his ways, especially the Cambridge coterie comprising Babbage, Baily, John Herschel, George Peacock (F.R.S. 1818) and William Whewell (F.R.S. 1821); they were to be outspoken critics of the Royal Society in 1830. But the Astronomical Society as such was to be no threat to the Royal, and in later years the two societies often worked harmoniously together.

The Council and its officers

Besides the quiet, even somnolent, weekly regular meetings of the Society there were some half a dozen meetings of Council every year, for even under such an autocrat as Banks the Council was necessary to the smooth running of the Society. For virtually all the Society's business, other than the reading of papers, was conducted by the Council, which dealt with both internal and external affairs, that is those which were properly the Society's concern and those in which it acted in response to government requests. Internal affairs concerned, first, the rights and obligations of the Fellows. There were requests for back numbers of the *Philosophical Transactions*, which then had to be collected from the Society's house; requests for postponement of admission; requests for copies of papers published in *Phil. Trans.*, or for their reprinting, or for the use of diagrams printed therein; requests for return of papers read but not printed (never granted, although papers in the archives might be copied); requests to withdraw or revise papers submitted before publication; requests to borrow books and instruments (permission depended both on the value of the item and the standing of the petitioner); permission for non-members to read in the Society's Library (always granted when the petitioner was known to some Fellow). The principal obligation of the Fellowship at large was the payment of dues, which in the early part of the century were still, as when the Society was founded, a shilling a week, with the possibility of compounding for a fixed sum at the time of admission; nevertheless members were often in arrears, and if more than two years elapsed without payment the Secretary had to be instructed to write to the erring Fellow threatening ejection.

Besides this routine business the Council had to consider and vote for the Croonian Lecturer (on the nature and laws of muscular motion),[12] the Bakerian Lecturer (by custom on a physical subject)[13] and the Fairchild Lecturer (a sermon).[14] The minutes do not record how those for whom the Council voted were chosen, but it seems probable that it was by Banks. There were also recipients of medals to be voted for: the Copley Medal (awarded normally annually since 1736 to the author, not necessarily British, of some important discovery or contribution), and the Rumford Medal, instituted by Count Rumford in 1796 for the most important discovery or 'improvement' in the study of heat or light (and extended to electricity) made during the preceding two years.[15] The Council formally accepted (and occasionally rejected) the offer of portraits and busts of distinguished Fellows. It voted on housekeeping matters, like the painting and repair of rooms and their arrangement, and the appointment of servants such as porters and cleaners.

The Council also voted for the Treasurers and Secretaries, at least technically, although Banks's choice was paramount; Presidents continued to regard nomination as their prerogative, although later in the century their choice was sometimes rejected. Vice-Presidents, on the other hand, were always nominated by the President, and one of these was almost invariably the Treasurer. This was later to create friction, for the Treasurer was usually chosen for his business acumen, and this was not often accompanied by scientific distinction. Banks's Treasurers were Samuel Wegg (F.R.S. 1753), an antiquary, replaced on his death by William Marsden (F.R.S. 1783), an orientalist and numismatist who served until 1810 when he was succeeded by Samuel Lysons (F.R.S. 1797), a barrister and antiquary, like Marsden a distinguished scholar; he in turn was succeeded by Davies Gilbert in 1819, the first Treasurer since the seventeenth century to possess other than antiquarian interests and one who was to play an extremely important rôle in the Society's affairs until 1830. Gilbert, born Davies Giddy in 1767 (he changed his name in 1814 in consequence of an inheritance) was the son of a Cornish curate of good family who made money by investment and he himself was to marry more money; he early showed mathematical talent which he seems to have pursued at Oxford as best he could; he also became friendly with Thomas Beddoes to whom he later recommended his protégé, the young Humphry Davy. By the 1790s he was active in Parliament, soon to become an M.P. in the Conservative interest. But he was also attracted to scientific pursuits: he was thrilled at meeting Banks in 1789 and enchanted at being elected F.R.S. in 1791; soon after this he was engaged with such Cornish engineers as Trevithick and

Jonathan Hornblower in the development of high-pressure steam engines, an activity which lasted for twenty years. He seems to have been an able enough Treasurer, but his intense conservatism and his inveterate and innate irresolution were to cause grave difficulties for both himself and the Society in later years.[16]

Of the Secretaries at least one was a scientist in these years: E. W. Gray (F.R.S. 1779) a physician and botanist, one of the scientific staff of the British Museum, was succeeded in 1807 by Humphry Davy (F.R.S. 1803), now successfully established at the Royal Institution, and he in 1812 by Taylor Combe (F.R.S. 1809), a numismatist on the staff of the British Museum, who edited the *Philosophical Transactions* from 1812 to 1824; Joseph Planta (F.R.S. 1774), a librarian at the British Museum who had been a Secretary since 1776 was succeeded by the distinguished scientist William Hyde Wollaston (F.R.S. 1793) in 1804 (so that for some years both Secretaries were chemists); Wollaston served until 1816, to be succeeded in turn by W. T. Brande (F.R.S. 1806), a chemist at the Royal Institution, who served for ten years. The Foreign Secretary from 1804 to 1830 was Thomas Young (F.R.S. 1794) who re-established the intellectual and administrative equality of the position with that of the other Secretaries. Thus of the three Secretaries in the first twenty years of the nineteenth century at least two were scientists in the modern sense. All the officers were hard-working, as shown by exemplary attendance at Council meetings; indeed Wollaston attended faithfully even after giving up his Secretaryship.

The Council also voted approval in financial matters, both such regular items as salaries, the expenditure of sums to be paid (usually to one of the Secretaries) for translation of foreign papers and for indexing the *Transactions*, and for exceptional items like charitable assistance to the families of clerks, Assistant Secretaries, porters and even cleaners on death or superannuation.[17] Although in this period the Treasurers' reports are not particularly business-like, they passed annual inspection by auditors (admittedly Fellows) as noted at the Anniversary Meeting each year. Indeed in 1800 it was found that the Treasurer had increased the Society's funds 'materially', so that after careful review of salaries it was concluded that they could safely be raised to keep pace with the loss in value of money since they had been fixed in 1743.[18]

Although the Council in this period does not on the surface appear to have concerned itself with the acceptance or rejection of papers for publication in the *Philosophical Transactions*, in fact it was entirely responsible for this important facet of the Society's public life. When the Society in 1752 took over the responsibility for publication of *Phil. Trans.* (for the previous sixty years it had been an official responsibility of one of the Secretaries, and before that a private enterprise of one of them) it established a Committee of Papers charged with all editorial decisions; and at this time, as for long, the Committee was composed of the entire Council, as the regulations revealed.[19] The Council could ask for advice, being 'at liberty to call in to their assistance' any member of the Society; although the Committee does not seem to have availed itself of the possibilities of refereeing very often in this period, it met regularly (usually half a dozen times a year between January and July) and kept careful minutes which show that all papers received serious consideration, and by no means all the papers read at meetings were thought worthy of publication. Those rejected were usually either irrelevant or trivial: the Society never took seriously the seemingly endless stream of English and foreign papers and letters offering solutions to such problems as the trisection of the angle, the quadrature (squaring) of the circle or perpetual motion, although some hopefuls wrote in the belief that the Society offered premiums for such solutions; nor did it accept naive accounts of 'monstrous births' as mere objects of curiosity, science, as it was thought, having outgrown such 'lusus naturae', and the science of teratology not having yet come into being.[20]

The Society as government adviser

There was thus a very reasonable amount of domestic business to occupy the Council, yet it composed what was only a very minor part of the Council's affairs, compared with the time and effort devoted to external affairs. For by 1800 the Society had been called upon from time to time for almost a century to advise the Crown and government departments, and this was to become more and more the case from 1800 onwards. Its most overt function of this sort was in connection with the Royal Observatory at Greenwich over which the Astronomer Royal presided, and to which since 1710 the Royal Society had acted as Visitors. Until 1830 the Board of Visitors, which inspected the Observatory yearly, consisted of the President of the Society and any Council members or Fellows whom he appointed. (In practice, the Astronomer Royal was always a member of the Council, so that by the early nineteenth century this was a matter of co-operation between the two bodies, rather than of control.) In fact, the Board of Visitors (as its reports to Council meetings show) served as an inter-mediary between the Astronomer Royal and the Board of Ordnance (and occasionally also the Admiralty) in matters arising from the need for new instruments and assistants, the payment of bills and salaries, and the printing of the Greenwich Observations compiled (not always with perfect regularity) under the supervision of the Astronomer Royal.

Initially linked with the Observatory, having been founded in 1767 by the then Astronomer Royal, Nevil Maskelyne, was the annual *Nautical Almanac*, intended to facilitate the determination of longitude at sea by the method of lunar distances, a method much used from the late eighteenth until well into the late nineteenth century. This involved the drawing up of tables of lunar motion for the coming year and exact determinations of the positions of selected fixed stars – all of which exactly conformed to the purpose for which the Observatory had been founded by Charles II in 1675. It also conformed to the aims of the Board of Longitude, founded in 1714 and put under the Admiralty; since 1718 this had included *ex officio* the President of the Royal Society, the Astronomer Royal, three Fellows of the Society, the Savilian Professors of Oxford and the Lucasian and Plumian Professors of Cambridge – a formidably professional group for the time. Maskelyne worked happily with the Board, whose interests he shared, but after he died in 1811 and was succeeded as Astronomer Royal by John Pond (F.R.S. 1807) things went less well. Pond was a keen and accurate observer, who laboured to improve the instrumentation at the Observatory and ultimately, in 1833, published a very accurate catalogue of fixed stars, but contemporaries thought that he lacked theoretical ability, and certainly he was not interested in the practical problems of navigation and left the *Nautical Almanac* to assistants without sufficient supervision or encouragement.[21] The situation is revealed in an exchange of letters between John Herschel and Babbage in 1814, when the young Herschel, still not certain whether to follow his father's profession, told his friend that he thought of applying for a post under Pond. Babbage wrote[22]

> I suppose you were not in earnest in what you said about the Nautical almanac. A less desireable employment you will not easily find out add to which, the situations are all occupied – It is an affair of *Patronage* in the hands of Pond, who (for what report saith) hath not proved the most eligible of Patrons.

This is perhaps not a very well reasoned criticism (Pond was, after all, technically responsible for the *Nautical Almanac*) and Babbage had already, with Herschel and Peacock, shown himself a fiery rebel against the Establishment, but the implications of the warning were almost certainly justified.[23]

So notorious were the deficiencies in the *Nautical Almanac* by 1818 that, apparently at the instance of Davies Gilbert, the matter was debated in Parliament, at which time John Croker (F.R.S. 1810, with Barrow a

secretary of the Admiralty, and often on the Society's Council) declared it
'a bye-word amongst the literate in Europe' that required greater accuracy
and closer cooperation with practising astronomers.[24] The Royal Society
could thus be seen as instigating reform; it was also to take a larger share
in the reformed administration, for the result of the Parliamentary debate
was the formation of a new commission to assist and advise the Board of
Longitude, now formally charged with the production of the *Nautical
Almanac*. Besides the President of the Royal Society, the Astronomer Royal
and the mathematical and astronomical professors at Oxford and Cambridge
(all F.R.S.) there were to be three commissioners resident in London and
three additional Fellows of the Royal Society: the first resident commissioners were Wollaston (no longer Secretary), Captain Henry Kater (F.R.S.
1814, an army engineer specialising in geodesy, soon to be on the Council)
and Thomas Young (Foreign Secretary). As critics were soon to insist these,
although all eminent scientists, were not specialist astronomers, but as
London residents and active men they, and not the sometimes indolent
academic astronomers who were there to advise them, were assigned most
of the blame, and their supposed defects were ascribed to their representing
the Royal Society.

It was not really surprising that it was the highly energetic and competent
Thomas Young, arguably the most distinguished among the commissioners,
who was appointed secretary of the Board, and superintendent of the
Nautical Almanac. He tackled the job with vigour and ably corrected
existing errors, but from the first he insisted that material of interest only
to professional astronomers had no place in the *Almanac*, which was
intended for sailors (and in any case he thought it extravagant to spend
government money on such things). Instead, he published such supplementary material under the title 'Nautical Collections' in the *Journal of the
Royal Institution*, to the annoyance of professional astronomers, who both
disliked having to look in two places for information and felt that national
prestige demanded that there be an exact British equivalent of the French
Connoissance du Temps. The *Nautical Almanac* was, however, only one
concern of the Board of Longitude, which particularly entertained and
endeavoured to promote every sort of proposal for improvements in
methods of determining longitude at sea. (It also acted as scientific adviser
to the Government when requested.) A notable example was the move to
establish an observatory at the Cape of Good Hope, a move which the Royal
Society actively supported. Thus in 1820 Davies Gilbert wrote of being
about to attend a meeting of the Board[25]

> When we have every reason to expect that the Establishment of an
> Observatory at the Cape of Good Hope will be formally

recommended to the Government and the first permanent introduction of Astronomy into the Southern Hemisphere may be considered as a commencement favourable to the Science of a New Reign.

The recommendation was indeed made, and the observatory established. But not even this success could silence criticism from astronomers, and agitation grew ever more intense from this period onwards, in a curious way reflecting on the Royal Society. For although the affairs of the Board of Longitude were peripheral to the interests of the Royal Society, the fact that its President was *ex officio* a member of the Board, and that so many commissioners were Fellows of the Society, inevitably linked Society and Board together in the public mind, for good or ill, and indeed the Board's business occupied a small but continuous amount of Council time.

Since the Board of Ordnance was well used to dealing with the Royal Society in connection with the Greenwich Observatory, it also turned to the Society for a certain amount of general scientific advice. In the 1790s the Society had assisted the Board in the planning and progress of the trigonometrical surveys then under way (especially in the matter of instruments), as well as advising on lightning conductors for magazines. In 1801 the Board of Ordnance asked for advice about the most suitable covering for the floors of powder works, to avoid danger of fire: as was already customary practice a committee was appointed, in this case consisting of Henry Cavendish, Charles Hatchett (F.R.S. 1797, not yet on the Council), with Rumford, Blagden and Gray being added later; all these except Gray, a Secretary of the Society, were experienced chemists, and Rumford knew a great deal about the practical aspects of gunpowder.[26] Two years later the Board of Ordnance sought advice about the cartridges used in cannon: these were normally wrapped in paper, and it had been suggested that various difficulties could be obviated by wrapping them in flannel – but how could the flannel be moth-proofed? On this occasion Banks showed a lack of judgement, for (perhaps thinking the matter urgent) he appointed as committee himself and the two Secretaries (Gray and Planta) – two botanists and a librarian, none of whom, as the report admitted, was at all knowledgeable about entomology.[27]

Fortunately the membership of most committees was more suitable. So when the Home Office asked advice in 1814 in connection with the storage of gas in Westminster by the Gas Light Company (there had been a dangerous explosion of stored gas two years before at Woolwich) the committee appointed included a chemist, Smithson Tennant (F.R.S. 1785, Council member 1813–14), and an engineer, John Rennie (F.R.S. 1798,

Council member 1808), as well as the ubiquitous Young; it met several times and was able to offer serious and constructive advice.[28]

Sometimes the Royal Society's advice was solicited by the Government in response to pressure in Parliament. Thus in 1816 the House of Commons 'prayed' the Prince Regent for an exact determination of the length of a pendulum beating seconds to be made at Greenwich Observatory and at the various stations of the trigonometrical survey begun in the previous century, the purpose of the determination being to arrive at an exact standard measure of length.[29] The Government promptly passed this request over to the Royal Society, and the Council responded by immediately appointing a committee including Davies Gilbert, Pond, Blagden, Young, Troughton (F.R.S. 1810, a notable instrument maker, Copley Medallist 1809) and Kater. The work was soon delegated to Kater, who was so diligent that the Council was able within a year to approve a report 'On the Basis of Linear Measure' for presentation to the House of Commons and to award the Copley Medal for 1817 to Kater for this work.[30] So impressive was his performance that when in 1818 he extended his geodesic work to Scotland the Society asked for, and obtained, considerable assistance for him from the Board of Ordnance.

In 1817 the Society on its own initiative actively endeavoured to influence the Government (really the Admiralty) to promote polar exploration; the report approved by the Council pointed out that polar voyages, besides providing geographical information, would give excellent opportunities for continuing the geodesic work which Kater had begun in 1817.[31] When the Admiralty did decide to send out two voyages during 1818 (part of the ill-fated efforts of the time to discover a North-West Passage) the Pendulum Committee resumed its meetings in order to plan experiments to be undertaken in the arctic regions.[32] Soon the Board of Ordnance was also involved, because the Society requested leave for Edward Sabine (F.R.S. 1818, at the time a Captain in the Engineers) to be allowed to sail on one of these voyages in order to perform the suggested experiments. In the event he accompanied Ross in 1818 and Parry in 1819–20, and thereby established such a high reputation for geodesy that he was subsequently sent on a voyage to the South Atlantic to determine accurately the shape of the earth (see Chapter 6).

Notable Fellows

During all this period the composition of the Council varied greatly from year to year. Officers, to their credit, all emulated Banks in faithful attendance – Blagden, who had ceased being Secretary in 1797, continued the habit of attendance for another twenty years. The Astronomer Royal Pond, like Maskelyne before him, usually attended, and was of course especially careful to be present when affairs of the Greenwich Observatory were discussed. Exceptional in attendance was Henry Cavendish (F.R.S. 1760) who, though reputed eccentric and unsociable, attended virtually every Council meeting until the month before his death (February 1810), a record nearly equalled by that of Samuel Goodenough, Bishop of Carlisle (F.R.S. 1789), a distinguished botanist who attended very regularly from 1809 through 1824, occasionally acting as Vice-President. Another occasional Vice-President was the notorious Sir Everard Home, successor to John Hunter whose work he plagiarised, fashionable surgeon, a Copley Medallist (1807) and Croonian Lecturer from 1817 to 1829, with only one interruption; his attendance in the last five years of Banks's Presidency was constantly good. Most of the Council came only as it suited them, although many of them were assiduous during the first months after their appointments; they included scientists, engineers, physicians, divines, antiquarians, naval officers, office holders (notably Croker and Barrow) and a copious number of noble lords, most of whom attended for a few meetings after first being appointed but seldom after, even when re-appointed.

A rough inspection suggests that Banks made sure that the Council was not a body which was in a position to challenge his authority; he wanted a body which would agree with him and decisions he could control, and he succeeded. It cannot, however, be said that the Society during this period failed to recognise scientific merit when, among other, lesser names one finds those of Wollaston, Davy, Troughton, Brodie, Ivory, Brewster, Kater and Seppings as awarded the Copley Medal, and Rumford, Leslie, Malus, Davy and Brewster the Rumford Medal.[33] But the atmosphere of patronage could not be dispelled, and younger men not unreasonably felt that the Royal Society under the aging Banks was too much entrenched in the antiquated world of the eighteenth century – a world which English society was clearly busily repudiating in all other walks of life. In 1819 there was visible some feeling for reform: on 18 March the Council ordered 250 copies of the charters and by-laws to be printed, and a copy of these in the Royal Society's Library contains handwritten suggestions for reform, notably relating to payment of dues and election of the Council. The 1820s saw agitation for reform in politics, law, religion, economics and social regulations. Inevitably there was such agitation as well in the Royal Society, not least because Banks's death in June 1820 suggested that repudiation of his attitudes and influence should be possible without inordinate upheaval.

Trial and error (1820–1830)

The problem of succession

Every reign draws sooner or later to a close, and a wise prince chooses his successor carefully and prepares him to rule. When the prince is an elected sovereign the matter is more complex, for he has not the full power of choice. Banks was in a difficult situation, for early in 1820 he celebrated his seventy-seventh birthday, and it must have been obvious that his long life was drawing to a close, although in spite of age and gout he continued to perform his duties by attending meetings of the Royal Society and the Board of Longitude. To all appearances he might continue to do so indefinitely. It was he himself who realised that the time was approaching when this would not be so, and there was need to choose a successor;[1] to this end he summoned a special Council meeting for 18 May 1820 at which, for he himself was not present, Sir Everard Home, acting both as a Vice-President and as Banks's personal medical attendant, announced to the very full meeting

> that they were called together by desire of the President who in
> consequence of the failure of his general health felt himself too
> much oppressed longer to perform the Duties of President, and had
> therefore with deepest regret requested Sir Everard to give the
> Council his resignation of that most honourable office which ha[d]
> constituted the pride and happiness of his life.

The three Vice-Presidents (Home, Gilbert and Goodenough) had known of Banks's intention for almost a month; to the other Council members it may have been unexpected. All certainly realised the dangers inherent in a forced succession: Banks had hinted often his personal opinions, and according to Gilbert[2] had recommended him to the other Vice-Presidents, while Home no doubt displayed his own ardent desire for the post. It must have been abundantly clear that Home would have been a disastrous choice, as his subsequent conduct was to show; the Society generally had disliked Banks's obvious favouritism towards Gilbert;[3] and in any case the Council members knew what was due to age and position. Promptly, tactfully, unanimously, the Council formally 'expressed the hope' that Banks would withdraw his resignation. He did so;[4] but only a few weeks later, on 19 June, he died and the choice of a successor was an almost immediate necessity.

In the five weeks between Banks's rejected resignation and the settlement of the question on 29 June a considerable number of candidates appeared. Among these were two royal candidates: Prince Leopold of Saxe-Coburg, later King of the Belgians, strongly favoured by his father-in-law George IV – but his position was weakened by Home, who discredited both the Prince and himself by spreading the false rumour that the King in fact did not approve[5] – and the Duke of Sussex, younger brother of George IV, a staunch liberal in politics and since 1816 President of the Society of Arts (but not yet F.R.S.). There was Lord Spencer (F.R.S. 1780), a frequent but inactive member of Council, whose interests were primarily literary, and the Duke of Somerset, who had, like Gilbert, been nominated to the presidency of the newly founded Astronomical Society and, again like Gilbert, had refused in courtesy to Banks.[6] Among Vice-Presidents (it should be noted that they were purely Presidential appointments) both Home and Gilbert were strong possibilities, the latter backed by Banks's approval. A very active candidate indeed was Sir Humphry Davy, Secretary 1807–12, brilliant professor at the Royal Institution, with scientific attainments far beyond those of Gilbert or Home, who hastened home from Paris soon after learning of Banks's resignation in order to canvass votes, which he did most actively, hoping for support from both the scientists and the fashionable world.

None of these candidates was very popular. Davy himself was by no means an attractive candidate to those who wanted a President like Banks, one who, as 'F.R.S.' wrote to *The Times*, was, like Banks,

> a man of high rank and acknowledged independence both in principle and fortune, a lover of learning and science and detached from all other pursuits... As science and learning are in a degree fashionable we may hope that some honourable and noble personage will step forward to claim our support.[7]

Davy may have thought of himself as conforming to these criteria, for he was, like Banks, a wealthy baronet; but others saw a world of difference between Banks, who came from the landed gentry, Eton and Oxford and was at home in Court circles, and Davy, an awkward Cornish lad educated

as an apothecary who had married money and, although he had achieved
both reputation and position, had failed to acquire the manner to go with
them. Nor was he favoured by those who wanted a scientific President of
a much more strictly scientific society. His scientific attainments and
eminence were not in doubt, but once again his manner was, especially his
lack of generosity to other scientists. As John Herschel put it,[8]

> The reasons for wishing that Davy should be opposed are
> grounded solely on his personal character, which is said to be
> arrogant in the extreme, and impatient of opposition in his
> scientific views, and likely, if power were placed in his hands to
> oppose rising merit in his own line, & not patronise it in others and
> in particular to involve the R.S. in controversies of much personal
> acrimony with the other learned European bodies.

So although Davy boasted (according to Herschel) that he represented both
'the interest' of the Royal Institution and 'the *aristocratical* interest'
neither scientists nor aristocrats were altogether eager to support him.

Indeed, there were other more eminent scientists in the Royal Society
who were wealthy (a necessary qualification because the President was
expected to entertain the Fellows frequently, as Banks had done) and
presentable. The young Cambridge graduates – Babbage, Herschel,
Peacock – saw such a candidate in William Hyde Wollaston (F.R.S. 1793),
who had been Secretary of the Society from 1804 to 1816, had served on
the Board of Longitude since 1818, a Cambridge man who had been a
practising physician and had made original contributions in physiology,
optics, crystallography and chemistry. He had received the Copley Medal
in 1802, had become a Fellow of the Geological Society in 1812, and was
a very rich man as a result of his discovery of a method for rendering
platinum malleable – perhaps the only serious thing to his discredit, for
some thought it discreditable to have made money out of scientific research
which should, they held, be solely the disinterested search for truth. The
chief disadvantage to Wollaston as a candidate was that he was socially a
retiring man, and was not at all eager to come forward. At first he declined
outright, but was persuaded by Kater (who had joined his supporters) on
the grounds that it was 'impudence' on the part of the Royal Institution
(Wollaston's own term) to claim to command the right of nomination of
officers of the Royal Society.[9] But he was never a keen contender, and
although older men than his original proposers soon joined the scheme,
canvassing letters were written,[10] visits paid, and all seemed favourable (the
other candidates seeming negligible opponents) in the end Wollaston

Fig. 2. W. H. Wollaston, M.D. (1766–1828), F.R.S. 1793, Secretary
1804–16, P.R.S. 1820.
By John Jackson

Fig. 3. Thomas Young, M.D. (1773–1829), F.R.S. 1794, Foreign
Secretary 1804–29.
By H. P. Briggs after Sir Thomas Lawrence

compromised, accepting the Presidency only until the Anniversary Meeting in November (it was widely believed that he could easily have won again then)[11] with the proviso that Davy should succeed him. His position, inevitably a disappointment to his backers, is understandable: the prospect of power held no appeal, nor the prospect of public life; he apparently did not greatly enjoy such social demands as general conversation at the Royal Society Club (though he attended its dinners) and while to his friends he was charming, he was too reserved to wish to dispense hospitality in Banks's manner;[12] above all, he was still anxious to continue his scientific work, and in this very year was actively engaged in electrical investigation in the wake of Oersted's epochal discovery.

Nevertheless, once in office he showed himself to be highly conscientious (he had already faithfully attended Council meetings for twenty years in various capacities, and was to continue to do so until his last illness in 1828). Indeed, he seems to have started the Society on the road to reform, even if in a minor way. Regular meetings continued much as before, but it is possibly no coincidence that eight papers were read on 29 June, probably the beginning of the custom of reading a dozen or more papers at a June meeting. The Council Minutes immediately reveal a more business-like approach than had been customary during Banks's later years. At the first meeting after the election (7 July) it was resolved that 'proper applications be made to persons in Arrears for Annual Contributions and that the Lists of the Defaulters be brought to the Council at the first meeting after Christmas', and no Fellow in arrears was any longer to be allowed the *Philosophical Transactions*: both quite reasonable enactments, although both produced a flood of complaints from Fellows who wanted the honour of being F.R.S. without the obligation. Wollaston summoned two Council meetings in November (16th and 23rd); besides routine business it was resolved to ask the Treasury for more space in Somerset House so that books and instruments might be properly housed; in response to an Admiralty query, a committee was appointed (consisting of Gilbert, Young and Kater) to report on the best method of measuring ships' tonnage; and a newly completed index to the *Philosophical Transactions* since 1780 was produced. No doubt Wollaston sighed with relief when his Presidential duties ended at the Council meeting on 30 November (to receive the auditors' reports) and the Anniversary Meeting, when the election was held.

Until the last minute, Davy was not sure of election, although by November opposition was no longer effective. Wollaston gave his support to Davy at the end of June and Gilbert also resigned his claims in favour of his protégé (generously, as Davy later acknowledged),[13] but by no means

all the influential Fellows accepted the idea of Davy as President. The
opposition's point of view is well represented by Sir Benjamin Hobhouse
(F.R.S. 1798), M.P. and barrister, who had written to Babbage in June

> In order to fill the vacancy made by the death of our most
> excellent, and much lamented President, Sir Joseph Banks, the
> Council of the Royal Society must choose out of it's [*sic*] own
> body, a President for the remainder of the year. When the annual
> election comes into the hands of the whole Society, which will be
> the 30th of next November, the Duke of Somerset, not now a
> member of the Council, will be proposed. His Grace, who has ever
> afforded liberal encouragement to science, possesses scientific
> knowledge and taste, and a fondness for the company of scientific
> men; in short, he possesses, in my opinion, all the qualifications,
> means, and dispositions, calculated to reflect credit upon the office
> of President; to advance the important objects of the Royal
> Society; and to secure the continuance of the respect in which it is
> held both at home and abroad.[14]

Somerset soon also withdrew, while Davy canvassed strenuously in early
July. In the event, his only formal opponent was Lord Colchester (F.R.S.
1793), former Speaker of the House of Commons and a barrister with no
scientific pretensions, who received few votes. Regrets continued; his friend
Johnstone told Somerset early in December that

> Davy does not seem popular: but the Society seems so much
> divided that I am glad you did not stand for the office of
> President.[15]

Davy's Presidency

So on 7 December 1820 Davy began seven years of giving an Anniversary
address 'On the Present State of [the Society] and on the Progress and
Prospects of Science'. He thereby established a tradition which now seems
highly appropriate. It was then not universally approved; the admittedly
prejudiced Johnstone reported

> Davy took the chair for the first time & opened the meeting with a
> discourse upon the present state of science & upon the best means
> of making further discoveries in each branch – his discourse was
> not a good one but evidently framed in imitation of Cuvier's on the
> anniversary of the French Academy;[16]

and he went on

> Instead of Sir Joseph's Sunday evenings, Sir Humphry is to have
> Wednesday evenings & some jealousy I understand prevails among
> the more humble class of our associates at the idea that Sir
> Humphry is to make a selection of those who are invited...

That this was not in fact the case is shown by the letters of Horner who,
on a visit to London from Edinburgh in 1821, remarked

> I have come home early with the intention of going to Sir
> Humphry Davy's this being the day on which his house is open to
> the weekly visits of such members of the Royal Society as will
> honour him by assembling at his house.[17]

Horner found Davy 'very accesible and civil' but, like Johnstone, he
thought Davy did not 'look' the President; rather, wearing semi-Court
dress 'in imitation of Sir Joseph', he looked 'mighty ridiculous' and 'very
like the porter at a shabby nobleman's gate'. (But it should be stressed
that Horner had favoured Wollaston.) Fortunately Davy himself was
happy, writing to a friend soon after the election

> I glory in being in the chair of the Royal Society, because I think it
> ought to be a reward of scientific labours, and not an appendage of
> rank or fortune; and because it will enable me to be useful in a
> higher degree in promoting the cause of science.[18]

And although many Fellows continued to disapprove, the scientific world
in general saw Davy as a successful scientist heading a Society which ought
to be composed of scientists.[19]

Indeed Davy did do a good deal to turn the Royal Society towards the
direction desired by the reforming scientists, away from the broad spectrum
of learning favoured by Banks, to confine its activities to 'science' as the
word was understood in English. In all of this he was greatly assisted by
the support of Wollaston, one of the most assiduous of Council members,
and it is to Davy's credit that, recognising his worth, he gave Wollaston a
prominent place on committees. A minor but significant indication of the
increasingly 'scientific' orientation of the Society can be detected in the
formal wording of certificates for candidates: slowly during the 1820s
literary and antiquarian interests disappear as the sole criterion of the
candidate's worth, to be replaced by, at least, devotion to 'literature and
natural knowledge, or science', or 'addiction to literature and science',

or 'attachment to science' when the candidate was not a practising scientist. But Davy did not control elections in the way that Banks had been able to do, as is shown by the election of Sir Francis Shuckburgh in 1824 even though Davy endorsed his certificate 'This certificate ought not to have been presented, there being no qualification mentioned' (it does not appear that he had any) and no new certificate was provided. Davy inherited his officers, and made no immediate change: Davies Gilbert continued as Treasurer and (by custom *ex officio*) Vice-President, Combe and Brande (of the Royal Institution) were Secretaries and Young was Foreign Secretary. In time there were changes: John Herschel replaced Combe in 1824 (he resigned in 1827), so that both Secretaries were scientists, and in 1826 J. G. Children (F.R.S. 1807, on the staff of the British Museum, who worked on both electricity and entomology) replaced Brande, to be a most conscientious Secretary.[20]

The Council also began to acquire a preponderance of men of science, although not without its share of noblemen and antiquaries. In 1825 the Council was disturbed because the Society of Literature, seeking a Royal Charter, sought to use the abbreviation for its members of F.R.S.L. which, it was feared, might confuse the two societies in the public eye (the Royal Society possessed the right of discussion of all such charters); for although the Royal Society might still elect literary men on occasion, it was in the process of repudiating its former inclusion of all learning, slowly narrowing its range to natural science.[21] But this was a tacit, not an explicit policy, for the Committee on the Revision of the Statutes for 1822–23 proposed only minor, though admittedly useful changes, which were duly accepted.[22] Over the years there developed a growing feeling that membership needed control, both qualitatively and quantitatively; some indication of this tendency is that contributions to the *Philosophical Transactions* (always noted in Presidential necrologies delivered at the Anniversary Meeting) were increasingly stressed. The feeling that Fellows who claimed to be practising scientists must be seen to be so is well conveyed by Whewell's remarks about submitting a paper: he told Herschel in 1823 that 'When I was admitted into the Royal Society [1820] I intended, if possible, to avoid belonging to the class of absolutely inactive Members', to which end he had 'been on the look' for a suitable paper: would the enclosed (on crystallography as a mathematical science) be suitable? Herschel enthusiastically endorsed the letter 'Yes. R.S. by all means', and the paper was published in 1825.[23] Davy encouraged promising contributors; for example he generously wrote to William Buckland (F.R.S. 1818, reader in geology, Oxford) in praise of his 'paper [which] was unanimously endorsed to be printed...I do not recollect a paper read at the Royal Society, which has raised so much interest as yours.'[24] (It received the Copley Medal for 1822.) It is perhaps significant that Davies Gilbert submitted a paper (on suspension bridges and the catenary) which achieved publication in the *Philosophical Transactions* for 1826; nearly twenty years earlier when he had read a paper on cogs he had not felt that publication was important to him,[25] but now, clearly, even he felt the need to display his scientific competence openly. The policy grew up, following Wollaston's lead, of having a number of papers read at the last meeting in June so that no great backlog of unread papers accumulated during the customary summer recess.

Agitation for reform by introducing more systematic qualifications than mere general interest continued, to become vociferous in 1827 when Davy, seriously ill since the previous autumn (although he had presided as usual at the Anniversary Meeting), was travelling on the Continent and the Society's affairs were left in the hands of Gilbert, Treasurer and Vice-President. On 1 March 1827 the Council

> resolved, that the best manner of limiting admission of Members into the Society be taken into consideration at the next Council, it being the opinion of the Council that the present mode leaves room for too indiscriminate an admission.

Later, the Assistant Secretary was asked to prepare an account of the number of members admitted yearly between 1800 and 1827, indicating those who had received medals or had contributed to the *Philosophical Transactions*. The committee surveying this information noted that the number of papers published by each member was fewer than it had been in the mid-eighteenth century, although more than in the early nineteenth century.[26] But it should be observed that no account was taken of the proportion of papers submitted to those published, nor of publication of papers in journals other than *Phil. Trans.*, such as those of the Astronomical and Geological Societies and the Royal Institution. In fact a fair number of papers languished without attention or came only slowly to the notice of the Committee of Papers.[27]

Some selection of candidates did occur. In 1822 Davies Gilbert had thought it worth noting in his diary that 'For the first time I was present at the Royal Society when a candidate was not elected',[28] but in fact blackballing was not totally unknown, although, under Banks, tactful Presidential interference had often caused withdrawal of doubtful candidates. Selection by blackballing was not without its own problems, for the backers of unsuccessful candidates were inclined to claim that injustice had been done. So acrimonious ill-feeling arose in 1826–27 when Babbage and James South, both highly excitable and argumentative, although normally friends, staged a violent argument over their respective candidates, leading Babbage to denounce the Council's conduct of elections.[29] Reform without ill-feeling was difficult to achieve, and the Council could in this case only insist upon the necessity of making the certificates conform strictly to the statutes. Actual limitation of numbers was an even more delicate problem. The principle was easy to state; the committee reported that

> The Committee assume, as indisputable propositions – that the utility of the Society is in direct proportion to its respectability – that its respectability can only be secured by its comprising men of high philosophic eminence. And that the obvious means of associating persons of this eminence will be the public conviction that to belong to the Society is an honour.

The principle was good; so too was the declaration that membership should be limited 'to such a number as could be a fair representation of the talent of the country'. But to implement such a principle was more difficult, especially when many influential members, and notably Davies Gilbert, feared all reform as dangerously democratic. It must be remembered that in 1827 conservative Tories in Parliament were actively and successfully opposing the liberal Whig demands for Parliamentary reform. Various important reforms were slowly to be enacted: the relaxation of the Corn Laws (1828) which lowered the price of bread, and Catholic emancipation (1829), both under the administration of the reactionary Duke of Wellington. But Parliamentary reform itself – the abolition of 'pocket boroughs' (controlled by individual landowners), the re-distribution of seats and the regularising of voting rights – had to await the advent of the Whig Government under Lord Grey (1830) and even then required two years of endeavour to secure its passage. Gilbert himself always sat for a pocket borough and lost his seat, together with his Parliamentary influence on behalf of science, in 1832. The Society was not more liberal than the Government, and in 1827, either because Gilbert was able to control matters or because a majority of the Council was also opposed to reform, nothing was done. The subject of limitation was not to be discussed again for some years, nor was limitation effected until 1847.

Improvement of the standard of papers read to the Society's meetings was an easier reform to attempt, although it too must have met with some resistance since it inevitably implied that the individual judgement of a Fellow was not enough to determine the suitability of a paper for presentation.

The judgement was, however, not at first very obvious, for ostensibly what was in question was only 'the order of reading of Papers at the Meetings of the Society', as the question (put by the Secretaries, Herschel and Children) was phrased in the Council Minutes for 26 April 1827. This was a novel problem and the Secretaries were asked themselves to devise some rules, to be approved by the Council on 3 May. The junior Secretary was to see that all papers were dated on receipt, their titles and dates entered into a book kept for that purpose into which was later to be entered the decision of a standing or sub-committee of papers consisting of the President, the Secretaries and two members of Council, elected annually. In case of rejection the junior Secretary was to send a printed letter informing the author that 'the Sub Committee, to whom the decision respecting the reading of Papers to the Royal Society is referred, have examined your Paper...and are of the opinion that it is not of a nature to be suited for public reading before the Society'; the author might then have his paper returned (papers read, even if not published, remained in the archives although the author might take a copy). It was also determined that published papers should carry the date of 'communication'; that the junior Secretary might, with the authorisation of the sub-committee, depart from the strict order of reception in reading papers; that he should prepare an abstract of each paper read, before submitting it to the Committee of Papers for a decision about publication; and that the President at each regular meeting of the Society should announce the titles of the papers to be read at the next meeting. All this was intended to regularise the running of the Society and improve the standard of the meetings. The first two Council members to serve on the sub-committee were Wollaston and Young, certainly among the ablest scientists and active over the whole range of both physical and biological science, although hardly representative of new blood.

The Society was, if only in minor ways, becoming more serious, and the officers ever more conscientious. The Secretaries now provided a fuller index to each volume of the *Philosophical Transactions* than had been the case earlier, and in 1822 began to index the Council Minutes. A Library Committee advised on the compilation of a Library Catalogue in 1821; by 25 July 1822 it was well advanced, consisting of an alphabetical catalogue of books and manuscripts.[30] This Committee also concerned itself with lighting for the Library: lighting meant that the Library could be available to Fellows for longer periods, and on 8 March 1821 the Council decided that in consequence it should be open in the evenings after the Society's weekly meetings (then held at 8 o'clock). As a result, the Library became a social centre, especially valuable for those Fellows not members of the limited Royal Society Club which met for dinner before the meetings. Soon tea was regularly served after the weekly meetings and intellectual as well as social ends must have been served.

It was an old tradition, dating from its foundation, that the Royal Society, being founded for the promotion of experimental learning, should possess instruments, to be used at meetings and to be employed by Fellows in their investigations. These were no longer needed for meetings, although they might be borrowed from time to time. Many lay neglected: in 1825 James South was to complain bitterly about the 'ruinous' condition of the equatorial which he had borrowed, and which he offered to repair if he might regard it as his own.[31] But instruments were still available for loan to those undertaking voyages for survey or geodesy, and it was not the fault of the instruments they borrowed that in 1826 Airy and Whewell were forced to abandon their course of experiments to determine the specific gravity of the earth by means of pendulum experiments at the bottom of a Cornish tin mine.[32] The Society kept a second's pendulum (regarded as a standard), and numerous meteorological instruments, the latter in regular use, with a Meteorological Committee to oversee the observations published annually in the *Philosophical Transactions*.[33] The Society had not yet recognised the changing needs of science, now advancing so rapidly as to demand totally new kinds of instruments, but at least it was beginning to take stock of what it had, along with other property.

An important and welcome development (though sadly productive of wrangles of far-reaching consequences) was the establishment late in 1825 of two Royal Medals, donations of George IV (and to be continued by subsequent sovereigns in turn). It was almost certainly the idea of Robert Peel, then Home Secretary, who had for some time been considering ways in which the Government could improve English science, especially zoology which he thought inferior to that of the French.[34] George IV was so favourably impressed by Peel's plan that he altered Peel's proposal of one medal worth twenty guineas to two medals worth fifty guineas each. King and minister left it to the Royal Society to frame suitable regulations; on 26 January 1826 the Council determined that the medals were to be awarded 'for the most important discoveries or series of investigations, completed and made known to the Royal Society in the year preceding the day of their award'. Unfortunately things did not go according to plan, and it was many years before the whole muddle was sorted out. In the first place, no physical medals were forthcoming: Chantrey and Sir Thomas Lawrence were each to design one side, but Lawrence died in 1830 with his share never completed, so that recipients did not receive their medals with the award. Then, when in 1826 the Council came to consider the matter, they decided – unwisely, since it was contrary to the regulations – to award the medals to two established scientists: John Dalton, who had indeed published a paper in the *Philosophical Transactions* in this year, but whose main work had been done long before, and James Ivory, a Cambridge mathematician of roughly the same age as Dalton, whom Laplace, it was said, thought one of the best geometers in Europe – but the award was for papers on mathematical astronomy of which the latest had been published three years earlier. Both were deserving of medals (Ivory had already received a Copley in 1814) but the flouting of their own regulations by the Council was, to say the least, injudicious. The already disaffected Charles Babbage (who had specially written a paper, read to the Society on 16 March 1826, 'On a Method of expressing by Signs the Action of Machinery', which he was sure deserved a medal) was to attack the action bitterly, and later to make it a basis for fierce criticism of the Society.[35] He and others spoke so contemptuously that in 1828 Ivory, hurt and puzzled, wrote to Herschel who had been Secretary at the time

> I now apply to you to be informed of the real truth about the
> Medal which was said to be awarded to me. What I have been told
> about it I can put no more reliance upon, on account of the cabals
> and intrigues with which everything relating to the Royal Society is
> mixed with, and which, for many years, have been a constant
> source of vexation and real injury to me although I have not been
> in the habit of attending their meetings.[36]

The regulations were promptly changed and the period altered to five years, but problems were to continue and the Royal Medal remained a source of vexation as well as pride to the Society for many years to come.

Babbage's complaint about the Royal Medal – and it was of course by no means the case that his was the only good paper published in 1826 – was the more ungracious because, when called upon by the Government, the Royal Society had consistently advised the spending of what became many thousands of pounds on the construction of Babbage's first calculating machine or difference engine through as usual a committee, which described the project as deserving of 'public encouragement'. It was not lack of support by the Society, nor of the spending of public money which caused the failure of the machine, but the fact that Babbage lost interest in his first design, wanting it to be replaced by a new, improved and even more expensive second design.[37] Ultimately the whole project came to a halt, and

in mid-century a workable, useful, and relatively inexpensive machine was
made by another (see Chapter 6).

The Council was frequently involved in astronomical matters, not only
as receiving the report of the Visitors of the Greenwich Observatory, but
in connection with the Board of Longitude and the *Nautical Almanac*. The
latter, under Young's superintendence, continued to fall short of what the
astronomers wanted. In 1822 South, an excellent observer, an ardent
member of the Astronomical Society, who was, however, to prove himself
irascible almost beyond the bounds of sanity when engaged in controversy,
published *Practical Observations on the Nautical Almanac*. This attacked
Young for the *Nautical Almanac*'s inaccuracy and its choice of contents,
comparing it most unfavourably with foreign productions. It was the first
round in a salvo from members of the Astronomical Society who were also
Fellows of the Royal Society; in the same year Francis Baily published two
attacks more temperate than South's but for much the same reasons. Young
stood firm, but was naturally hurt by what he saw as the 'injustice and
impropriety' of South's attack, for he thought that as there were other
journals in which such information could be published (as he himself
published 'supplementary' information in the *Journal of the Royal Institution*) comparison with the Continent where there were fewer journals (and
professional astronomers, being in secure posts, were content to undertake
routine scientific activities) was quite irrelevant.[38] This was not, strictly
speaking, Royal Society business, but so many Fellows were either *ex officio*
on the Board of Longitude or appointed to it, and criticism came so much
from Fellows, that all this was necessarily of concern. It was to prove,
indeed, part of the later public attack on the Society for permitting science
to languish in England. And in 1824 the Council was directly involved by
the Board of Longitude's application for advice about

> what additions ought to be made at the public Expense to the
> Nautical Almanac for the use of practical Astronomers in general,
> without any view to their immediate utility for the purposes of
> Navigation.[39]

Astronomy at this period was a subject much open to debate and criticism
of a most acrimonious kind. As Visitors of the Royal Observatory the
Society came under attack in 1825, when Stephen Lee (F.R.S. 1798),
Assistant Secretary since 1810, provoked a violent storm by virulent
criticism of the Greenwich Observations for 1821 recently published under
the supervision of the Astronomer Royal, John Pond. Very properly his
letter was first referred to Pond, and then a competent Royal Society

committee was set up to examine both criticism and reply. The committee decided that Pond had not been 'culpable', but had been negligent and allowed many errors to creep in; shortly after this the Council tried in vain to get Pond to appoint some 'superior assistants' with astronomical competence instead of the 'drudges' he preferred. Lee was asked to apologise for the manner of his attack, but preferred to resign.[40] Other astronomical problems were easier: asked by the Home Secretary for advice about measuring an arc of meridian in New South Wales, or by the Admiralty for an astronomer to superintend the observatory recently established at the Cape of Good Hope, the Council found no difficulty.[41]

Besides astronomical matters the Council was asked by the Government to advise on a wide range of problems; in each case a committee was appointed, its report discussed by the Council, and a letter embodying it sent to the government department concerned. Thus in 1822 the Admiralty called for an opinion about the belief that the use of coal tar in place of pitch to protect ships' timbers was the cause of illness among the crew; in 1823 the navy demanded advice about lightning conductors, and the Court of Common Council for re-building London Bridge about granites to be used in its construction; in 1824 the Customs wanted to know how to remove the danger of contagion from imported silk or cotton without damaging the fabrics; in 1825 Peel asked for a committee on gas-works, now becoming widespread. There were frequent queries about scientific matters to be investigated on the voyages of exploration that were such a feature of this period – in 1822 there was a committee 'to draw up a series of Queries upon scientific subjects for the use of such Travellers as may be desirous of extending the boundaries of natural knowledge' – but each voyage now brought its own problems and while in 1821 pendulum and longitude investigations were of particular interest, soon systematic geomagnetic studies became more widespread (cf. Chapter 8).

Two government-inspired investigations stand out as of scientific importance, even though both were practical failures. The first began in February 1823 when the Navy Office asked for help in improving the quality of the copper sheathing on ships, designed to prevent insect attack and fouling of the timbers but itself very subject to corrosion. This was a subject very much to Davy's taste, and he took it upon himself to investigate the matter personally, borrowing one of the Society's balances 'for prosecuting his Experiments on the corrosion of copper'.[42] He found that corrosion of copper could be prevented by fixing pieces of more electropositive metals to it; trials at sea proved this to be the case, but unfortunately it also rendered the copper ineffective as an anti-foulant. Davy published several excellent papers, including the Bakerian Lecture for 1826 'On the Relations of Electrical and Chemical Changes', but in practical terms the investigation had been unsuccessful. It also produced a great deal of rancour, both because the committee resented his having taken the matter into his own hands and because when he published his results others claimed to have made the same discovery earlier. The other investigation of this kind was also to prove of little practical use, and to provoke criticism.[43] This was the prolonged study (the committee set up on 6 May 1824 and continued until 1831) with the aim of discovering how to improve glass for optical purposes. Of the active sub-committee of Faraday (F.R.S. 1824, in spite of Davy's opposition), George Dollond (F.R.S. 1819) and Herschel, Faraday was to analyse chemically the glass in use and to devise experimentally a better glass, Dollond was to act as practical optician and make lenses from the improved glass, and Herschel was to test its optical properties. Funds were supplied by the Government, much encouragement was given by the Board of Longitude, hoping for better glass for astronomical uses and for light-houses, and Faraday worked very diligently to create a special heavy glass. Although Dollond as a practical optician did not find his samples useful for lenses, it was with a specimen of this glass that Faraday discovered diamagnetism in 1845, and the idea of heavy glass was to be taken up later in the century for spectroscopic work. The investigation, like Davy's, promoted natural knowledge (after all the purpose for which the Royal Society had been established) even though, again like Davy's, its results were of no direct practical use.

The election of 1827

Clearly good work was being done under the Society's aegis, and many good men were being elected to the Fellowship. But on the whole the period of Davy's Presidency was not a happy time. The abortive attempts at reform begun early in 1827 were undertaken behind Davy's back, as it were, for Davy suffered a stroke late in 1826 (the first of several), and in the new year was advised to travel abroad to a drier climate. He left Davies Gilbert, Treasurer and Vice-President, in charge. Gilbert as already noted opposed all reform, whether in the political life of the country or in the Royal Society, and the failure of the reforming party may have been due mainly to his political acumen, for he was chairman of most of the relevant committees. Aware that Davy must soon resign, Gilbert was also exercising his political skill in finding a suitable successor as President, although he managed to keep this from public knowledge. Six months after leaving England Davy

Fig. 4. Sir Humphry Davy, Bart (1778–1829), F.R.S. 1803, Secretary
1807–12, P.R.S. 1820–27.
By Sir Thomas Lawrence

Fig. 5. Davies Gilbert (1767–1839), F.R.S. 1791, Treasurer 1819–27,
P.R.S. 1827–30.
By Thomas Phillips, 1833

wrote to Gilbert (on 1 July 1827) resigning the Presidency from the Anniversary Meeting, stressing that ill health (not opposition) was the cause of his resignation, and – aware that much of the opposition arose from his insistence that the President had the right to choose both his officers (he had appointed Children as Secretary, rejecting Babbage whom some at least favoured) and his successor – only hinting at the problem of succession. After describing his poor state of health he wrote to Gilbert

> Under these circumstances, I feel it would be highly imprudent and perhaps fatal for me, to return, and to attempt to perform the official duties of President of the Royal Society. And as I had no other feeling for that high and honourable situation, except the hope of being useful to society [*sic*] so I would not keep it a moment without the security of being able to devote myself to the labour and attention it demands. I beg therefore you will be so good as to communicate my resignation to the Council and to the Society at their first meeting in November, after the long vacation; stating the circumstances of my severe and continued illness as the cause...
>
> It was my intention to have said nothing on the subject of my successor. I will support by all the means in my power the person that the leading members of the Society shall place in the chair; but I cannot resist an expression of satisfaction in the hope you held out, that an illustrious friend of the Society, illustrious from his talents, his former situation, and, I may say, his late conduct, is likely to be my successor... I wish my name to be in the next Council.[44]

Clearly Gilbert had been looking for a suitable successor for some months, probably in fact since Davy's first illness, for the Duke of Sussex (a not very likely candidate in 1820) was in 1826 and 1827 in correspondence with his librarian and medical adviser T. J. Pettigrew (F.R.S. 1 February 1827) about the possibilities of the office, although he was not as yet strongly inclined to seek it.[45] Gilbert did not apparently initially seek the office himself, although he was better suited to it in 1827 than in 1820: he had been a reliable Treasurer and committee man in the Society (as in Parliament); he had helped promote the Government financing of Babbage's difference engine; he had continued his work on Cornish steam engines and was regarded as an authority by the Institute of Civil Engineers in 1825; he had been active in bridge design, which led to his 1826 *Phil. Trans.* paper; he had been approached as possible President of the Astronomical Society again in 1823 (in declining he explained, as earlier, 'I was not fully convinced that the Royal Society would not be displeased by seeing their Treasurer & Vice President nominally at the hand of another Society'); and although not an eminent or distinguished scientist, he was a competent mathematician and had made some real contributions to practical and theoretical engineering.[46] Everyone spoke of his pleasant manner and amiability, although he was thought too pliable and indecisive. Perhaps on this occasion, as later, he hoped to be both king-maker and a power behind the throne. At any rate, having accepted the interim Presidency from 6 November 1827 (when Davy's letter was read to the first Council meeting since the summer recess), Gilbert spent his time in a frantic attempt to promote the candidature of his choice, for which he had Davy's approval, that of Robert Peel, whom he had first approached on 13 February 1827.[47]

It was an attractive idea. Peel was intelligent, forceful, conservative, principled, a patron of science – F.R.S. 1822 and the instigator of the Royal Medals – and, since April 1827, out of office. He was attracted by the proposal, especially when Gilbert, in describing to him the Council meeting of 6 November, assured him that he need expect not 'the least difficulty' at the Anniversary election. So far so good. But in the next three weeks Gilbert wrote a stream of letters to Peel, giving details of statements and gossip by himself and others, raising doubts, suggesting problems, then being reassuring again; although he constantly told Peel that he could not fail to win the election, he also constantly told him of acts of opposition. Gilbert professed to be more worried about the choice of Secretaries, yet he told Peel of every suggestion, possible and impossible, for the Presidency. Somerset had been spoken of, and Lansdowne, and on the 15th Gilbert told Peel that

> Doctor Wollaston suddenly [*sic*] brought forward the subject of who should be President. I expressed my astonishment at having that matter mooted. Doctor Wollaston declared that he had never consented & that he would put up Mr. Warburton (the Member for Bridport)

whose election, Gilbert thought, would alter the character of the Society. It is not possible to say how serious Wollaston was in suggesting his close friend Warburton (F.R.S. 1809), who had made his money as a timber-merchant, was a radical M.P., a leader in the attempt to reform the College of Surgeons and currently involved in attempts to pass the Anatomy Bill (to legalise the sale of cadavers for dissection), nor how seriously Gilbert took the threat. Peel was reassured by Thomas Amyot (historian and antiquary, F.R.S. 1824) who was his keen supporter, but Amyot declared that a more serious threat might come from Gilbert himself, whose friends were trying to persuade him to stand; however there was, he thought, 'no possibility of his yielding to them' and soon reported that Gilbert 'has declared he will not act, if elected', although Gilbert himself, oddly it might seem, was at the same time asking Peel's advice as to whether he (Gilbert) should resign if elected. No wonder that Peel, in the face of Gilbert's manoeuvring, coupled with newspaper speculations, decided on the 21st to withdraw his name, even though Amyot and Barrow urged him to continue since only 'the democratic party' was opposed to him, while Gilbert continued to insist that he wanted Peel as President. But with no other candidate remaining Gilbert's election was inevitable.

It is difficult to make sense of all this unless Gilbert, for all his agitated and protesting wavering, had not aimed at the Presidency all along. For there was no concerted opposition to Peel – Warburton differed from Peel in politics, but had no pretensions to being a scientist; few thought that Wollaston would be a candidate again, and indeed he expressed himself satisfied with Gilbert; and no other scientific name was proposed. Gilbert was far better qualified scientifically speaking than anyone else, he was thoroughly familiar with the Society's affairs and conscientious in the performance of its duties. It is difficult to accept Barrow's judgement that Gilbert was 'a mere child in the affairs of the world', too naive to understand what was happening; in view of his very considerable political experience it is far more likely either that he had always, consciously or unconsciously, wanted the Presidency, and worked to get it without seeming to do so, or that he drew back from his support for Peel because he learned (perhaps through Pettigrew) that the Duke of Sussex was still interested in the Presidency, and he knew that time was needed to prepare the way.[48] In the event, it was to take three years before that time was ripe.

A second interim 1827–30

Meanwhile Gilbert was President, and it seemed that few or none of the Fellows felt particularly strongly about it. To all appearances mild, he would not be expected to domineer or even dominate, inclined as he was said to be to agree with everyone. True, he had been successful in his choice of Secretaries: Herschel had resigned (he disliked working under Gilbert) to be replaced by P. M. Roget (F.R.S. 1815 for his invention of a slide rule, M.D., a practising physician, who was currently investigating London's water supply for the Government), and Edward Sabine (F.R.S.

1818, an artillery officer on leave, who had received the Copley Medal for geodetic work and recently had worked with Babbage on terrestrial magnetism) replaced Children, although Faraday had been seriously proposed, and South had proposed himself. But Gilbert had lost his desire to continue as Treasurer, a move he had made on the grounds that he would probably resign after a year – even a friendly Council saw the impropriety of this. Instead, Kater (in 1820 an active Wollaston supporter) was chosen, putting in a position of power a professional scientist, even though a retired army officer, and a friend to the reformers. In spite of Gilbert's well-known conservatism there was some hope of reform now that he was President and no longer acting for Davy.

Indeed Gilbert was too experienced a committee man not to tighten up and improve the administration of the Society. In this he was particularly aided by the industry and methodical mind of the senior Secretary, Roget, who rapidly gave his heart to the Society's administration, although inevitably in the course of twenty-one years of service he sometimes forgot that he was only the Council's servant, and that it was necessary for him not only to follow the procedure resolved upon by the Council but to be seen to do so. Thus for the first time the Council Minutes of the first meeting after the Anniversary record the various committees then in existence, the Council's resolution to continue them, their membership and any reports presented by them. The Sub-committee for Papers, which decided on those papers suitable for reading at the meetings, continued, and it is not clear whether it was approval or disapproval of its activities which caused the Council (13 December 1827) to resolve 'That the Papers proposed' be placed in the Society's rooms 'for inspection and perusal by the Members of the Council, at least one week before the meeting of the Committee', which provided some check upon the Sub-committee, albeit a very loose one.[49] On 11 December 1828 the Sub-committee membership, besides the officers, consisted of Wollaston and Young. By 1829, Wollaston and Young having died, the Council members were Thomas Bell (F.R.S. 1828, surgeon and zoologist), Robert Brown (F.R.S. 1811, botanist), Kater and Warburton; since the Secretaries were Roget and Sabine (the latter replaced by Children in 1830) it was heavily weighted towards the biological side, but not at all subservient to Gilbert. It is perhaps significant that in the great burst of criticism which erupted in 1830 there was little criticism of the papers read, although much of the traditional manner of their reading with no opportunity for discussion.

Gilbert did achieve some control over the Council, most notably over the possibility of limiting membership: when the report of the committee was read it was ordered to be printed for distribution to the Council, but after brief consideration (14 February 1828) it was allowed to drop. This tactical victory by Gilbert was never forgiven him by would-be reformers. In revenge, after the Duke of Sussex was elected a Fellow on Gilbert's proposal, the Council approved an amendment to the statutes, ending the privilege of princes of the blood royal, peers and privy councillors to be elected on the day of their proposal, and requiring the elapse of a month between proposal and voting. Ordinary Fellows were to be made more active, at least in the payment of subscriptions. The Secretary was given rules for dealing with Fellows in arrears, normally the result of carelessness. Fellows were often indignant at being expected to pay promptly; although most paid after prodding, some few allowed their Fellowship to lapse. Those who found the fees too dear (£10 for admission, and either £4 a year or a composition of £40, as distinct from the original, long-continued shilling a week) did not seek admission, in general.[50]

Much other domestic business was discussed. There was a Committee on 'the most eligible mode of printing the *Philosophical Transactions*' at the beginning of 1828, but it was found that the Fellows disliked any change in format. The Council set up a great number of meetings of various financial committees from early November 1827; it is difficult to tell whether this reflects on Gilbert's efficiency as former Treasurer, or was an attempt by the Council to ensure that the new Treasurer, Kater, was properly efficient.[51] The property owned by the Society, both real estate and that housed in the Society's rooms, was closely scrutinised. A sensible departure was the offering to the British Museum in 1828 of the Arundel Manuscripts (a seventeenth-century presentation), a transfer effected in 1830, and of the Jones Collection (an eighteenth-century presentation) to the India Office, a transfer effected a little later. By modern standards this was a most proper thing to do, especially as the British Museum paid for the gift, partly in duplicates, partly in money to be spent on the purchase of scientific books, and the Howard family, who had given the manuscripts, approved. At the time critics saw the action as a cause of censure, claiming that the Society should have sold both collections on the open market at a (presumptively) greater profit.

The most imaginative and far-seeing event of 1828 was the creation by Wollaston of the Donation Fund, to be applied after his death 'in promoting experimental researches, or in rewarding those by whom such researches may have been made'. He urged the Council to spend the income 'liberally', with the wise stipulation that no 'benefaction' should be made to a serving Council member. The fund was promptly augmented by various Council members (further augmented over the years it exists still), the Council gratefully recognising what Young, the senior officer present, shyly described as 'the way in which Science ought to be encouraged in this country, not by tormenting the Government to do this, that and the other for us, but in doing what is wanted for ourselves, which is the truly dignified character of an independent English Gentleman'.[52] In this they were more percipient than most contemporary Fellows, who sadly failed to make much use of the fund for over twenty years, finding it difficult to think in terms of 'research grants' for personal use. Much more comprehensible and desirable must have seemed Wollaston's simultaneous gift of a sizable quantity of platinum and palladium, to be given to Fellows on application for use in suitable chemical experiments. But the Donation Fund, a splendid memorial both to Wollaston's generosity and to his vision, was to be much used in the latter half of the century.

Interest in the Society's activities was increasing, and reputable journals such as the *Philosophical Magazine* (one of whose editors was now a Fellow) printed accounts of papers read at meetings regularly, until (28 February 1828) the Council resolved that no copy was to be taken of any part of the minutes of the Society's proceedings by any individual, although the Secretaries were authorised to communicate accounts of the proceedings 'at their discretion'. This was intended partly to stop indiscretions in newspapers, partly to maintain discreet silence about committee reports on the work of individuals, partly to stop certain Fellows, notably Babbage and South, from obtaining long extracts from the minutes or publishing confidential material. It was a reasonable resolution, but the need for it was an ominous foreshadowing of future acrimony, for it gave both Babbage and South the opportunity to argue that there must be a discreditable reason for such secrecy.

But 1828 was not a tranquil year, and it is not surprising that as the autumn approached Gilbert considered resigning in favour of the Duke of Sussex, whom he approached through Pettigrew. The Duke was receptive, but cautious, writing

> If proposed, I shall certainly accept of it, and do my best to
> forward their interest... I should imagine they would wish to get
> me on the Council first and if so I should not object to such
> nomination if it does not call me into London too frequently.[53]

But having made the proposal Gilbert took fright and, finding that the Council members he consulted disapproved of a royal President, repudiated his suggestion in cowardly fashion.[54] Soon there followed a sad dispute over the award of the Royal Medals. At the Council meeting of 20 November 1828 it was resolved to give a Royal Medal to Encke (for his work on comets); some wished to give the other to Wollaston for a paper to be read that day as the Bakerian Lecture for the year ('On a Method of rendering Platina malleable'); while Kater and Fitton proposed the name of Mitscherlich for his discoveries in crystallography. The Council members were unfamiliar with Mitscherlich's work, but they appointed a committee, of which Herschel was a member; this reported that he was indeed worthy of a medal, then or in the future. The Council, however, voted to give the medal to Wollaston – as Young told Herschel, 'This may be the last opportunity of paying him a *just* compliment', pointing out further that the report on Mitscherlich would be just as useful the next year (as indeed it was). Herschel had evidently expressed the opinion that the Council's collective ignorance of foreign and superior science showed that English science was in decline, for Young added

> I do not understand your *forebodings*...I do not know the precise nature of your feelings with respect to the R.S. but I am fully ready to admit with you that much is *nicknamed* science that has little pretension, to the name of any thing more than slight of hand at best. And I can willingly admit that great injustice is done to men of science both in public and in private, as I suppose *every one* thinks he has experienced. My own inclination is to go on for the short remainder of my life doing what good I can and to die in harness, without caring for any man's approbation but my own, if I am able; and I should trust that you partook of the same spirit.[55]

Young's feelings of gloom are understandable, and his resolution the more commendable, for not only was he facing continued attacks on his *Nautical Almanac* from individuals (notably South and Baily) but in July 1828 Parliament had abolished the Board of Longitude. On 26 June Croker, Barrow's senior colleague at the Admiralty, obtained leave to bring in a Bill 'to repeal the Acts that constituted the Board of Longitude', assuring the House that the result would be a saving in public money without any loss of efficiency. When questioned by Sir Joseph Yorke, himself a former Lord of the Admiralty (1810–18), Croker insisted that

> the purposes for which the Board of Longitude was originally established, either having been already effected or found to be altogether impracticable [it] was unnecessary to continue an establishment so constituted any longer, and that the purposes

which it still continues to effect, might be better and more
advantageously executed by the Council of the Royal Society

(of which he was a Fellow). At the Bill's third reading (4 July) Davies
Gilbert was careful to point out that the proposed abolition of the Board
of Longitude, which he regretted, was on the grounds of economy, not
because of any 'misconduct on the part of its members' and stressed the
importance of the Admiralty's consulting the Royal Society on 'all
questions connected with the science of navigation'. Croker agreed
heartily, regarding the Royal Society as a much more suitable body than
the Board for consultation and disingenuously emphasising that the
Nautical Almanac would continue, since it was edited 'by a gentleman who
happened to be Secretary of the Board' but not by the Board itself.

Young, who had not been forewarned, was overcome by what he called
'the *sudden* demolition of our Board'; he noted sadly, 'I knew mischief was
brewing, but I could not have believed it to be so near at hand'.[56] He was
now left without official advisers; as he told Herschel, asking for advice,

> I am told that the Bill leaves me Superintendent of the N.A.
> without any new appointment and I must therefore proceed to do
> whatever is urgent on my own responsibility: there is scarcely a
> Board of Admiralty to direct me. And how I am to be directed
> hereafter I know not: but the first thing I have to do is to make up
> my own mind as to what is best, with the advice of those friends
> who will be charitable enough to assist me.

Herschel was willing to advise Young personally, but, as a gesture of
independence, he pettishly refused to accept the fees due to him as a member
of the Board of Longitude during 1828, claiming that non-cooperation was
the way to show the Government what scientists thought of its actions –
although he had in fact intended to resign from the Board of whose activities
he disapproved, like other astronomers. He also saw its relation to Royal
Society affairs. As he wrote to Kater

> What *will* the civilized world say to the cavalier kind of way in
> which science and men of science are treated in England!...
> I hope and trust that *if* the business of the Board of [Longitude]
> *is* to be thrust upon the Council of the RS that body will at once
> adopt and keep up the *highest tone.* At once refuse to attend to *any*
> *questions of small moment.* – Never give the slightest ground for
> suspicion of a job. – And what they *do* recommend take care to
> have executed to the letter. But by what right does Government

presume to lay this duty on our shoulders. Our situation is irksome enough and responsible enough as it is, and a mere *recommending* body is in the eyes of many, a more useful tool to bear responsibility – in simple words – a catspaw![57]

The Admiralty in fact had decided in October to set up 'a resident committee' (that is one whose members were resident in London) to be composed of three Council members of the Royal Society, each at a salary of £100 a year; the committee was to advise 'on all questions of discoveries, inventions, calculations and other scientific subjects'. (The Royal Society pointed out that difficulties might arise because Council membership was a yearly status, but the Admiralty foresaw no problems arising from annual change of membership.)[58] The first three commissioners were Young, Sabine and Faraday (the Admiralty wanted a chemist, without giving any reason, so Young suggested Faraday). Baily, South and the Astronomical Society generally were quick to note the absence of astronomers, and promptly pointed out both this and defects in the *Nautical Almanac*, including the absence of much astronomical data which they wanted (but did not yet wish to supply themselves). Young's view was that

> Mr. Baily will never rest satisfied until the Astronomical Society not content with the humiliation of the Royal Society shall succeed in dictating to the *Admiralty* and the *British Parliament*, and the warfare begun in the Morning Chronicle is no doubt to be continued in the House of Commons, that is if the attack of 'banditti' unresisted because contemptible, can be called a *warfare*.[59]

Clearly Young's superintendence was incompatible with the presence on the new committee of Fellows of the Astronomical Society. When Young died at the beginning of April 1829, the Astronomer Royal (Pond) was once again placed in charge of the *Nautical Almanac*, although he was not much interested in it. Finally after the Astronomical Society secured its Royal Charter (1831) it was recognised as being a possible adviser to the Government. For the first time the Royal Society was not the only representative of British science in Government eyes. In the wake of William IV's succession to the throne the Royal Astronomical Society joined the Royal Society in providing Visitors of the Greenwich Observatory, and took over supervision of the *Nautical Almanac*. The two Societies quickly learned to work in harmony, and the unedifying spectacle of Fellows of the Royal Society attacking their own institution ceased, although not without major warfare which Young did not live to see.

The Royal Society's Council had indeed a great deal of astronomical and related work to oversee: besides the Board of Longitude and Greenwich Observatory there was the continued work of the Pendulum Committee and of the Optical Glass Committee as well as consideration of the reduction of astronomical data supplied by the two southern hemisphere observatories at the Cape and in New South Wales and of the desirability of maintaining these two observatories at government expense. There was also the question of whether the Government should indeed continue to finance the extremely costly and troublesome construction of Babbage's calculating machine, as well as to advise how much time and money should be expended on the trigonometric survey of Ireland and on the survey of the Thames for the proposed 'levelling of its banks'.

It is curious that problems of physical science should have been so dominant when the proportion of physicians and surgeons was so high among the membership (100 in 1830 out of a total membership of 662). But physical science was not only flourishing in the 1820s but had reached a stage when it required funding; it was also potentially useful. Biological science was not in this state, and agriculture, once so prominently placed among the Society's concerns and still a sufficient qualification for membership, did not call for support at this time. Further, many medical men, like Young or Wollaston, were frequently, even if in medical practice, more interested in physical than in biological science. On the other hand, the presence of so many medical men in the Royal Society brought the Society within the range of attack, at a time when the medical world was in turmoil. It must be remembered that the 1820s saw the attempt to gain status for apothecaries, and to reform the College of Surgeons (1830), while Thomas Wakley, the fire-eating editor of *The Lancet*, attacked all privilege in the medical world, and by extension in the Royal Society. Little or none of this strife, fortunately, appeared in the Society's day to day affairs. Wakley's approval of some Fellows of the Royal Society, like Warburton or Charles Bell, in no way prevented him from attacking the Society: it was a privileged body, there were, as he saw it, abuses of privilege, and individuals of whom he disapproved were elected into it. The first attack in 1829 was over the election to the Society of Bransby Cooper, a surgeon at Guy's Hospital who had recently bungled an apparently simple operation; perhaps less significant than the attack is the fact that Wakley had an ally within the Society, for he both knew the names of those who had signed Cooper's certificate (these included Sir Henry Halford, president of the College of Physicians, Astley Cooper and Babington) and that he was elected with a majority of one (no doubt the Council had *The Lancet* among other papers in mind in endeavouring to maintain the confidentiality of Council Minutes). Worse was a letter in June 1829 ambiguously signed 'H' and significantly indexed as 'Royal Society, jobbing at', which not only attacked the Society for proposing to elect Cooper, but asserted that the medical profession was making the Society 'a snug nest' (was this a gibe at Roget, the first medical Secretary for ten years?), thereby converting 'an institution founded on the most liberal principles, into an engine for party purposes and self-interest'. Further, 'H' declared, evidently not entirely *au fait* with trends within the Society, that as Warburton had been promised the Presidency on Gilbert's resignation as a reward for bringing in the Anatomy Bill 'it now remains to see if the promise will be fulfilled' when the Bill had failed.[60] It was splendid material for Wakley, who never relaxed in his attacks over the next twenty years.

The crisis of 1830

Although the election of 1829 went off smoothly to all appearances, with the officers unchanged, there was clearly much discontent; private criticism among the Fellows was becoming more and more public, soon to make the attacks on the *Nautical Almanac* seem mild. Discontent among the astronomers over Young's attitude, and the failure of both the Royal Society and the Government to enforce reform, tended to make them declare that English science was in decline, and that this sad state of affairs in which scientific merit was unrewarded by the Government and there was no incentive for hard work was the fault of the Royal Society. So Herschel could lament privately in 1828 that it was idle

> to attempt competition with our Continental neighbours whether French or German in matters of science generally. Our day is fast going by, and as we are both proud and poor and negligent we are rapidly dropping behind in the race,[57]

a theme he was later to pursue publicly (in the article on 'Sound' in the *Encyclopaedia Metropolitana*), claiming specifically that English chemists had failed to pursue problems raised by Continental chemists, but had been beguiled by Oersted's discoveries to play with galvanism instead of working at chemistry (possibly an attack on Faraday for what is now regarded as his most important work).[61] Young's reply to Herschel was very much to the point, and in the older tradition of English science. He wrote

> I fully agree with you that we are poor and proud as a country – and too negligent of other nations: but I hope you are

mistaken that our day is fast going by: for I do not apprehend that
our scientific reputation has ever depended on the caprice of a
ministry or of its agents. What had King James to do with
Newton's Principia? Or George the 3rd with the analysis of the
alkalis [by Davy]? I do not mean that it would not have been more
honourable to them if they had: but I do not know that it would
have been much the better for astronomy or for chemistry, or for
the scientific reputation of this country, which as far as I can judge
from my late intercourse with foreigners, stands very nearly where
it has always stood since my recollection. You will go on with your
astronomy as best suits your taste: I should have more leisure for
my optics and my hieroglyphics if I were to be no longer an
almanac maker under the high *patronage* which I now enjoy and for
which I am *properly* grateful: the act [abolishing the Board of
Longitude] is disgraceful to the government but its effects I trust
will be overcome by the spirit of individualism[.][62]

Yet although Young appealed to an established English tradition of
independence in science, he was in some ways more representative of a
new outlook than were Herschel and his friends. Young was, by any standards, a professional scientist, depending on salaried posts ('places' in
the pejorative language of the times) whose duties he carried out conscientiously. At the same time he regarded scientific research as a private
matter to be financed, if at all, by private, not public, endowment. Herschel,
Baily, Babbage and South were all independently wealthy, and lacked any
real sense of public responsibility: Herschel was pleased to be able to withdraw from advisory work on the *Nautical Almanac* which took up time, and
none of them saw anything reprehensible in Babbage's twelve years as
Lucasian Professor of Mathematics at Cambridge with total neglect of all
duties, although they complained of Oxford and Cambridge professors
remote from London as members of the Board of Longitude (in fact most
were regular attendants, being given travelling expenses). They attacked
Young for continuing to cling to his posts, and to the Foreign Secretaryship of the Royal Society,[63] but at the same time they campaigned for the
creation of more posts and more government patronage, not realising the
responsibilities these entailed. It is difficult not to believe that they wanted
rewards without obligations, rewards such as the knighthood given to South
in 1830 when he threatened to emigrate to France. They were also very
ready to allow paid posts to lesser men, but they all, regrettably, seemed
to have begrudged the professionalism of some of the best scientists of the
day, like Young and Faraday. The Cassandras announcing the decline of

science in England thought of themselves as advocates of the professionalisation of science, and to a certain extent they advanced this cause, but it
was chiefly to advance by means very different from those they advocated.

One of the first public statements of the position that it was the Royal
Society which was responsible for the decline of science since the days of
Newton and of Davy was in a letter signed 'F.R.S.' appearing in *The Lancet*
on 3 April 1830. F.R.S. wrote that

> the glory of our Society is fast fading away, and must soon cease to
> be, unless the members exercise their latent energies, and rescue
> the noblest institution of our country...from the degrading
> condition to which it is now reduced – that of a medical advertising
> office, a very puff shop for the chaff of medical scribblers, where
> you may see prospectuses of patent medical inventions, treatises on
> the treatment of diseases, and systematic works on anatomy &c.,
> written either by members or candidates for that honour...

of whom the writer thought there were too many. (Presumably the books
which offended him were presents, and the prospectuses book announce-
ments.) This would not be worth noting, were it not that there was worse
to come. For at the very end of April 1830[64] there appeared Charles
Babbage's *Reflections on the Decline of Science in England*, which was to
create a great storm reflecting gravely on the Royal Society.

Since Babbage's book was the focus for the concerted attack of various
Fellows on the Society to which they had all sought election, it is important
to consider his arguments in some detail, since in the main they are typical
of what his associates thought, though these lacked Babbage's personal sense
of grievance. His main position (and to this Herschel and David Brewster
(F.R.S. 1815) publicly subscribed, while more did so privately) was that
the exclusive reason for the decline of science which they all felt existed was
the position of the Royal Society which, unlike its French counterpart, was
neither exclusively scientific nor a government department. (Babbage was
by no means alone in failing to comprehend the difference between the
Institut, which included academics devoted to all branches of learning, and
the Académie des Sciences, confined to the natural sciences, or to realise
that the Royal Society over the centuries had often approximated more
closely to the former than to the latter – had it remained unchanged the
British Academy would not have been needed.) That the Royal Society's
membership should be limited to practising 'scientists' only – the very
word was only slowly coming into usage at this time – without the Royal
Society's traditional admixture of scientific patrons (many of them noble-

men), antiquaries, 'lovers of learning', barristers and politicians was a fair point of view, one to which very many of the younger Fellows subscribed. The problem which would then arise of how to finance the running of the Society was one to which even those like Babbage himself, or Herschel, or Baily who had been on the Council paid no attention, and in this they were possibly right, for other societies, like the Astronomical or the Geological, were apparently solvent (but paradoxically *their* membership was not limited to practising scientists). Nor did they see that government support meant to some extent government control. It is a measure of the fortunate differences between France and England that the Royal Society could become more strictly scientific without government support or control.

Babbage's other grounds for attack, set forth in great, even libellous, detail, were more personal, more debatable and in some cases trivial. He attacked the leadership of the President on all fronts: while it was fair to complain that the reforms proposed by the committee for limiting membership[65] had been allowed to drop, as indeed would (probably) not have been the case had the Fellows at large had more control of the Society's affairs, Babbage at the same time criticised Gilbert as being too mild and amiable to control his Council. He attacked all pluralism. He disliked all *ex officio* posts, like that of the President on the Board of Trustees of the British Museum or on the erstwhile Board of Longitude. He attacked the appointment as officers of the Society of men with salaried posts elsewhere. (It must be remembered that he himself had wanted the Secretaryship when Davy gave it to Children, who was on the staff of the British Museum, and that Brande, a previous Secretary, was in the employ of the Royal Institution.) He criticised the appointment to the now defunct Board of Longitude of advisers holding other, paid posts (notably Young); and the appointment in any capacity of army officers (he had no regard for Sabine's scientific attainments). He attacked the appointment of medical men (of whom he thought there were too many, besides insisting that their papers were not really scientific, while, he claimed, they brought what he called the jealousies of their profession into the Royal Society with them). In fact as a physical scientist he was narrow in his definition of science, and as a rich man he lacked any conception of true professionalisation, distrusting all those who earned a living by science, which was the avocation of such men as himself and his close friends. In our terms Young, Brande, Children, Roget, Kater, Faraday, even Sabine are professional scientists, and men like Babbage, Herschel, Baily, South or Fitton, who were wealthy by inheritance or marriage or retired after making a comfortable fortune to devote themselves to science, are professional only in the sense that the seventeenth-century founders of the Royal Society were professional; but Babbage did not see it so.

These were complaints of principle. There followed a flood of criticism in apparently uncontrolled, even petty anger. It was the fault of the Society that Sabine's work contained errors, which Babbage specified at length. He complained of the Secretary's refusal to send him a free copy of the Greenwich Observations (he was told that they were given only to institutions, private individuals being expected to buy them). He criticised the Society for aiding 'private' individuals to obtain government support (but not, of course, its continued recommendation that the Government should spend thousands of pounds on his own calculating machine, support which it would never have received without Gilbert's activity over many years). He complained about every aspect of the award of the Royal Medals, not least the failure of the Society to award him one in 1826. He attacked the Council for not allowing the Fellowship absolutely free access to the Council Minutes, and Roget for what now seem some very minor defects in the Council Minutes (which Babbage was of course able to study).[66] He complained, with justice, that the Croonian Lectureship had been virtually monopolised by Home, and with less justice about the Fairchild Lectureship, a non-scientific sermon with a trivial endowment, which it was not always easy to find preachers to undertake. He complained both that the qualifications of candidates were not scrutinised with sufficient care, and that a candidate proposed by himself had nearly been rejected because the certificate was not in proper form. As he told Herschel in advance of publication

> I hope to teach even chartered and ancient bodies a lesson that may in future prevent them from studiously neglecting and then insulting any individual amongst them and with the aid of public opinion I will make them writhe if they do not reform. In short, my volume will be a receipt in full for the amount of injurie I have received and like a bill once paid I shall forget them and my anger together.[67]

Like others before and since, Babbage, easily offended himself, believed that he could be as rude to others as he liked because his motives were good; he did not feel that they should be offended or reply in kind.[68]

Herschel, agreeing as he did with many of the criticisms, yet disapproved of Babbage's method of attack, although, from friendship and lack of political acumen or involvement, he hardly saw the whole danger of it. Before the publication of the book he advised sensibly

As for your book by all means burn it – whatever be its contents &
whatever good you may expect it to do – it was written under the
influence of feelings & prospects so different from those which I
trust will brighten the rest of your life that you will not read it with
much satisfaction hereafter;

while after publication he wrote prudently

I have read your Book which will I think do good and would have
done much more had it been less bitterly sarcastic, and had you not
been so sweeping in the attribution of low and mean
motives...This is an age of reforms...I hope and trust that no false
steps on the part of the Royal Society will prevent it from
benefitting as much as possible from the present conjuncture.[69]

Public reaction varied, on the whole aghast at the affair. It was from other
Fellows that significant comment emerged. Those who agreed with Babbage
praised the book to the skies, notably Brewster who described it promptly
and enthusiastically in the *Quarterly Review*. Others were more cautious,
thinking, like Dalton, that it could do good if it stimulated the Royal Society
to greater efforts, since clearly some of Babbage's 'strictures' were justified.
Others again, like J. F. Daniell (F.R.S. 1814), rejected Babbage's major
premise, pointing out that a country which had produced Young, Wollaston
and Davy could not have been long in decline, a position with which
Faraday evidently agreed since he was responsible for the publication of a
defence of English science by his friend Gerrit Moll of Utrecht.[70]
    The Royal Society began by ignoring the book officially, although
rumours of impending censure by the Council – carefully conveyed to
Babbage by his friends – flew about. Gilbert, temperately, wrote to query
the accuracy of Babbage's facts.[71] Roget (presumably) persuaded Gilbert
to defend him publicly from the charge of falsifying Council Minutes; hence
at the end of May (and not before a *Times* leader had called on the Society
to refute Babbage's charges if possible) the President and Secretary wrote
to the *Philosophical Magazine* attempting to explain and justify their actions
(by stating the facts) in a letter picked up by other journals and newspapers.[72]
Then at the Council meeting of 10 June 1830

Captain Kater[73] stated that he wished to take the liberty of
proposing a question upon a subject in which he conceived the
dignity and well being of the Society were concerned. As the
Charter invests the Council with the sole government of the Royal
Society, and the exclusive management of all its concerns, he
conceived that one of the first duties of the Council was that of

preserving the Statutes inviolate, and of noticing any infringement on them. He therefore requested to be informed whether any, and what, steps were intended to be taken respecting a publication by Mr. Babbage, entitled 'On the Decline of Science in England'.

The President thereupon observed, that, deeply as he regretted the injurious tendency of Mr. Babbage's publication and disapproved of the uncandid spirit which pervaded it; and notwithstanding the violations of the Statutes, which had, in strictness, subjected its author to the penalty of ejection from the Society, he was yet unwilling in consideration of the past services which Mr. Babbage had rendered to science, to proceed to this extremity: but thought it would be more consistent with the dignity of the Society to waive all further notice of it.

No doubt Gilbert was sensible to refrain from letting Babbage appear a martyr; and Kater, having done his duty, let the matter drop. Not everyone thought this a wise move: on 8 July *The Times* published a letter from 'F.R.S.' which, after a detailed account of the Council meeting, cited the section of the statutes which would have justified Babbage's expulsion, namely, that he had 'by speaking, writing or printing, publicly defamed the Society'. Babbage was unrepentant; not content with having escaped censure he brashly complained about the content of the Council minute, inasmuch as it recorded conversation, not formal business. Gilbert tried to soothe him, protesting ruefully 'My most earnest endeavour through the whole of my life has been to please and gratify everyone – a fruitless endeavour as I have found in numerous circumstances.'[74] Babbage then, politely this time, wrote to ask that the offending Council minute be 'expunged'; this was not done, although on 11 November it was 'rescinded' on the grounds that it contained no resolution; after this, on 18 November, Roget had to be cleared of blame.

All the publicity in the summer caused great agitation. Sir James South wrote to Babbage as result of the letter in *The Times* to try to get up a 'party' beginning with himself, Baily and Babbage, to institute reform, as he saw it, in the Society, a move which only branded the astronomers as dissidents more firmly than ever.[75] The Society's official reaction to the letter in *The Times* was silence. But, meeting on the same day to approve a Humble Address to the new King, William IV, the Council voted 'That the Minutes made at the Meetings of the Council be not open to the general inspection of the Fellows of the Society before they have been read and approved at the next meeting of the Council', a wise provision which, if enacted earlier, might have saved the Council much trouble.

The election of 1830

The Society adjourned for the vacation, not to meet again until 4 November 1830. But the possibility of Gilbert's resignation was obvious,[76] as indeed the 'public press' quickly saw. On 14 August the *London Literary Gazette* reported rumours of new royal patronage for the Society which it congratulated on the grounds that this would be 'likely to quash all the cabals and disagreements which have for some time unfortunately interrupted the purer pursuits of scientific knowledge'. This patently inspired rumour presumably came from Pettigrew, acting as the Duke of Sussex's political agent in the as yet unreformed Parliamentary spirit. In his earliest surviving letter of this period (10 September) Gilbert could claim that he had not initiated the campaign and 'still' had doubts lest the Duke was too socially exalted for the position, though he declared that he himself would be pleased and honoured to see the Royal Society 'under the protection of one so near the throne', ambiguously adding that he would welcome the Duke either as his successor, or as a member of the Council. Thus when, four days later, Gilbert wrote to the Council to deny the rumour that he intended to resign he must have realised that his hand had been forced and he must resign, whether he liked it or not. When Pettigrew sent Gilbert's letter of the 10th to Sussex, the reply was unambiguous acceptance: 'The Presidency of the Royal Society is a situation which I shall feel proud in filling, but I would not have accepted of it without learning in the first instance whether such a Situation would be pleasant in the highest Quarter'.[77] But having found that it was, he was prepared to accept if elected, and showed himself anxious to take the position seriously. No wonder that by the 26th Gilbert was promising to send another circular letter to the Council announcing His Highness's 'intention', and was deep in planning the future Council, while soon Pettigrew was enquiring whether Gilbert would not like to be Treasurer again since Kater was anxious, so it was said, to relinquish the post. Clearly Pettigrew and Gilbert expected an easy and smooth transition to a new regime.

So when Gilbert at last wrote his letter to the Council on 1 October (without suggesting an early Council meeting) he did not expect any violent opposition. In this he was wrong. A great deal of sentiment within the Council, as within the Society, was opposed to the idea that the President could choose and appoint his successor, as had been apparent in earlier elections. Many were opposed to the idea of a royal prince as President, as Gilbert well knew, fearing either that he would be too exalted and autocratic, or that he would be a mere figurehead, with Gilbert the power behind the chair. Supporters came either from those who genuinely welcomed the idea or from those who thought it impossible to propose an opponent to royalty, even while extreme opponents looked for a rival candidate. The Duke himself showed quiet good sense, writing to Pettigrew that he intended to stay out of London for a time, adding 'I could say nothing at present relative to the Royal Society...I have everything to learn and I do not want to commit myself before I am master of the subject.'

Meanwhile, inevitably, rumours were flying and the day after Gilbert's letter to the Council the *Literary Gazette* could report pretty accurately that

> the differences of opinion which have arisen in the bosom of the Royal Society are far from having been reconciled. The minutes of the Council are still impugned by Mr. Babbage, in a way likely to lead to more paper war; and the question of the Presidency is also a subject of not very pleasant discussion. After some overtures had been made to His Royal Highness the Duke of Sussex, and conditionally accepted by him, we understand that a change of sentiment sprung up, and that a number of the Council desire to retain Mr. Davies Gilbert in the Chair.

What with this, and similar remarks in the *Spectator*, it is not surprising that Pettigrew in alarm asked Gilbert to come out in the open; he responded by sending a copy of his letter to the Council, assuring Pettigrew that he would be flattered and honoured to assist the Duke in any way possible.

Once again, inspired accounts made the whole affair public. The *Literary Gazette* for 16 October 1830 reviewed the whole story, publishing Gilbert's letter to the Council and remarking

> It is understood that, under a prince of the blood royal and so near the throne, *there will not only be no impropriety, but a positive advantage, in the President's resuming his station as Vice-President and Treasurer.*

One cannot but wonder whether Gilbert was not the source of this comment, either directly or indirectly, for the *Literary Gazette* continued (obviously with an eye to Babbage)

> The Fellows, generally, as far as we have met with those to whom this prospective course was known, express themselves much pleased with it; though in so numerous a body, with its division of astronomers and mathematicians, there may probably be a small minority of dissentients.

And the article continued to express approval, both of the proposed officers
and of 'probable' modifications in the charter, with new by-laws about the
control of papers read and published, the quality of proposed members and
the possibility of division into classes, either into two (scientific and
non-scientific Fellows) or, like the Institut, into divisions of the sciences.
Ten days later this optimism was contradicted in a letter to *The Times* signed
'F.R.S.', but the editor of the *Literary Gazette* promptly reported that he
was sure of his facts because he had seen letters exchanged between Gilbert,
Pettigrew and Sussex. So if Gilbert was not the source, Pettigrew must have
been. In any case, Pettigrew promptly wrote to contradict 'F.R.S.', and to
insist that there was no wish 'to force His Royal Highness on the Society';
on the contrary, he had been approached two years previously to serve on
the Council, which he had thought 'infra dig', but he was nonetheless a
genuine Fellow, not an 'honorary' one. (It will be recalled that Sussex had
then also feared being 'merely at the head of a party', which he was anxious
to avoid.)

At last, in this uneasy atmosphere, the Council met on 4 November; after
routine business there was read a letter signed by thirty-three Fellows
(some on the Council) demanding that the President supply information
concerning 'the nature and particulars of any correspondence that may have
taken place' in regard to proposed changes in the officers of the Society.
It was immediately resolved that the signatories be invited to meet the
President and Council 'to consider the proper measures to be taken in the
present situation of the Society' – a blow indeed to Gilbert's principles. The
meeting duly took place on 11 November, when the letters exchanged
between Gilbert and Pettigrew were laid on the table for all to read. At the
end of the meeting, Roget, as senior Secretary (and one can imagine in some
embarrassment), enquired what should be done with the letters; it was
resolved that they should be returned to the President without being
entered in the minutes, presumably to avoid embarrassment all round.

Sussex himself was in an ambiguous position: he was interested, he was
even anxious to be President, and saw possibilities for the kind of liberal
reform that he had always favoured, but he was inevitably far too royal to
accept opposition easily, and he had no wish to be a mere party candidate,
for he rightly foresaw internal dissension as dangerous for the Society's
well-being. And he was dependent on Gilbert who had so often changed
his mind in the past about resigning, and who still wavered or at best hoped
to retain power without responsibility. Neither Sussex nor Gilbert can have
been pleased by the outcome of the meeting on 11 November: not only were
secret negotiations brought out into the open, but the meeting had passed

a resolution stating that it was 'of the highest importance' that the officers and Council 'should be elected from among such members of the Society as are, by their acquaintance with the conditions and interests of science, best qualified to discharge such offices': qualifications which Sussex could hardly be said to meet. Moreover, in defiance of Gilbert, Herschel and Faraday proposed that it be recommended that the President and Council should provide not just a house-list of officers and Council but also a list of fifty Fellows out of whom the Society at large should choose the officers and Council.

Meanwhile on the same day there was published another attack on Gilbert and the Society, in the shape of a pamphlet by James South entitled *Charges against the President and Council of the Royal Society*.[78] South claimed that this was 'a brief outline' of a work he had already announced 'On the Conduct of the Royal Society as connected with the Decline of Science in England, together with Arguments, proving, that before the Society can regain Confidence at Home, or Respect from Abroad, a Reform of its Conduct, and a Remodelling of its Charter, are indispensable'. The pamphlet is a wild farrago of charges, little likely to achieve its aim of influencing votes at the Anniversary Meeting. South complained that the minutes were both incorrect and inaccessible (he had been refused permission to see the minutes before they had been approved at the next Council meeting); that the Society had not been good Visitors of the Greenwich Observatory; that the Society had spent money wrongly, the Society's on Anniversary dinners and on gilding and cleaning pictures, the Government's on such projects as optical glass and a measurement of levels of the Thames; that it had lost some of its property, without knowing how or when; that it had refused to discuss the report on the limitation of membership; that it had awarded medals unwisely, and improperly, out of ignorance and carelessness; that it had kept Ivory off the Council in favour of a relative of the President; that it had 'permitted' an officer to canvass for votes for the Council; that it had entered a threat in the minutes against 'one of the *best* friends of science and of the Society' (Babbage); that when Sussex declared his wish to be on the Council the request was refused as 'improper'. Though South could conclude 'Personal animosity I disclaim' it is difficult to see what else motivated him. It is not surprising that Gilbert refused on the Society's behalf to accept a copy of the pamphlet as a present, although his action was injudicious. It is not surprising either that an anonymous 'Lover of Science' should have produced a pamphlet entitled 'The Knighthood of Sir J. South A Death Blow to Science', insisting that although the knighthood (really intended to keep South from emigrating to France as he threatened to do) was said to be awarded for scientific achievements and 'liberality and public spirit in the cause of science', his achievements were not particularly great, his private fortune was spent on his own observatory only, and he had become 'the leader of a discontented party', exciting 'dissensions fatal to the Society which was instituted for the fostering of science, which he professes himself so ardently desirous to promote'.

This was comedy, and South could do no real harm, except to strengthen the belief that it was the astronomers who were the chief malcontents. More serious, and perhaps more worrying because it came from a medical man, was *Science without A Head, or the Royal Society Dissected by One of the 687 F.R.S. – – – sss*, in fact by Augustus Bozzi Granville, as he was to acknowledge in the second edition (1836, *The Royal Society in the XIXth Century*). Granville, an Italian with an Italian M.D., who had settled in England in 1813 after political activity at home and subsequent service as surgeon to the English fleet, had studied midwifery and chemistry in France, was F.R.S. 1817, and a highly successful London physician, who knew far more about Continental science than did most of his English contemporaries. He fiercely denied Babbage's claims that English science was in decline, or that English education had made it so (he later could not resist a reference to Babbage's dereliction of duty as Lucasian Professor); he thought poorly of Brewster's review of Babbage's *Decline of Science*, especially its inaccurate appraisal of the state of Continental science; and he doubted South's wild charges about the Royal Society's misuse of money. Science in England, indeed science in London, he thought flourished. But the Royal Society did not and reform was needed. He adopted Babbage's 'dissection' of the membership into those who had published papers in the *Philosophical Transactions* and those who had not, and he, like Babbage, discussed the problem of lectureships and medals. (Of Babbage's claim to a Royal Medal he commented wryly that he was sure that his 'labours' might entitle him to one, but certainly not to the exclusion of those who had been the actual recipients.)

He proposed two areas for reform: in the general management of the Society, and in the election of the officers. He depicted the Society as a kind of Venetian republic or oligarchy, with the Council more or less continuous at the whim of the President; he deplored the dullness of the meetings, at which much formal business was read out, the paper was frequently interrupted by announcement of the results of voting for candidates, while the audience slumbered, aware that they would not be allowed to discuss the paper being read. He believed that the Royal Society should be divided into scientific classes, with a small 'Free class' for persons interested in science but not practising it, the whole being limited in number. As for the choice of officers and especially President, since at present the Society was 'without a HEAD', Gilbert being, although 'clever and amiable', too indecisive, he favoured Sussex, not only as a potential head, but as being likely to restore the control for which Granville had admired Banks. Who else, Granville asked, was there as candidate? And indeed no one had yet come forward. According to Granville there were rumours that Warburton, Herschel or South would make good Presidents, but to this he retorted,

> Now, although the former of these gentlemen knows a *good deal*, and the second deals in *infinitesimals*, and *there is no dealing* with the third

none of these were qualifications for a President.

Truly at this point there seemed no other candidates and the Council devoted itself both to routine business and, nearly acceding to the wishes of the meetings of 11 November, to drawing up a house-list of eleven present Council members to be retained, and twenty-nine non-Council members out of whom ten new members should be elected, the lists to be printed and sent to all Fellows living within the range of the twopenny post (near London) and at the English universities. At meetings on 23 and 25 November the award of medals was decided upon, and on 30 November the auditors' report was received, after which the Council adjourned for the all-important Anniversary Meeting with its election. Nothing was said about these excitements at the Society's weekly meetings (18 and 25 November); rather curiously 18 November was the date chosen for the reading of a sound but somewhat old-fashioned mathematical paper by Gilbert (on imaginary and negative numbers).

Meanwhile, as the Anniversary day approached, two serious candidates emerged: the Duke of Sussex and John Herschel. Herschel had for some time disliked Gilbert's behaviour in the Royal Society. His resignation from the Secretaryship in November 1827, ostensibly because he wanted to devote more time to science, probably stemmed really from his reluctance to serve under Gilbert. In 1830 Gilbert, charged as President of the Royal Society under the will of the Duke of Bridgewater to find one or more persons to write a work or works illustrating 'the Power, Wisdom and Goodness of God as manifested in the Creation' consulted the Archbishop of Canterbury and the Bishop of London, who advised having eight authors. Gilbert asked Herschel to write on astronomy, but he loftily declined, saying that he did not choose to write for money and instructing Gilbert that he should have given the opportunity to poor men who lived by science; in contrast Babbage, offended at *not* having been asked and angered at the apparent emphasis on biological science, nominated himself to write a ninth treatise, a fragmentary work on mathematics and religion.[79]

Herschel was highly indignant at the manner of Gilbert's formal resignation, writing[80]

> if Mr Gilbert has neither the capacity, nor the spirit to hold his station with dignity, he at least ought not to be suffered to imagine that he can hand it over like a rotten borough to any successor be his rank and station what they may.

On the other hand he does not seem to have intended to put himself forward as a candidate; rather it was his friends, Granville's 'noisy ones', who began canvassing for him late in November. The first advertisement in *The Times* (also issued separately)[81] was signed by sixty-three supporters. That Herschel was not really very anxious for the honour is apparent from his telling Babbage

> this measure has placed me in the very difficult predicament of either refusing to undertake a duty to which I am called by a large number of respectable men, many of them my personal friends, or of incurring *the most distressing* interruption and very possible ultimate grievous injury to those objects I had made up my mind to regard as the main business of my scientific life.[82]

And he added that if he were to be elected he would serve one year only. But ignoring this, in the short time left his friends canvassed actively, arguing that only by the election of a scientist as President could the scientific character of the Royal Society be saved.

On St Andrew's Day, at the Anniversary Meeting of the Society, the Council was elected first, with mixed results: some of those favoured by Sussex's party were elected, but so were some of those favoured by Herschel's party. Herschel was not elected to the Council, so that technically he was not eligible for the Presidency, but this was apparently overlooked. The voting was close: 119 for Sussex, 111 for Herschel (approximately a third of the Society had voted). For Secretaries, Roget and Children where chosen, for Treasurer, John Lubbock.

So Sussex won, in the contested election he had hoped to avoid. The reasons for this are various. Sussex had many ardent supporters: some, like Granville, knew his ability; others, like Granville again, thought that, as Banks had been a better President than Davy, so a royal duke would be a

Fig. 6. John Frederick William Herschel (1792–1871), F.R.S. 1813,
Secretary 1824–27.
By Christian A. Jensen, 1843

Fig. 7. H.R.H. Augustus Frederick, Duke of Sussex (1773–1843),
F.R.S. 1828, P.R.S. 1830–38.
By Thomas Phillips

better choice than any scientist; some thought it improper that he should
be opposed at all, when he had indicated his desire for the post. (So the
London Literary Gazette (27 November 1830) commented

> We do not remember an instance of any branch of the royal family
> being opposed in a similar manner as this setting up another
> candidate against His Majesty's brother,

while a letter in *The Times* 'regretted' that a son of William Herschel, who
had received royal patronage, should oppose a son of George III.) These
views were understandable; even today the situation might pose difficulties.
Others genuinely failed to see in Herschel the truly eminent scientist his
friends proclaimed him. He was still relatively young (thirty-eight in 1830)
and had a fairly small scientific output, his *Treatise on Astronomy* embodying
his survey of the southern celestial hemisphere appearing only at this time,
so that Granville was not being grossly unfair when he wrote of Herschel
that he

> has written half-a-dozen papers on mathematics and astronomy,
> and has the further merit not only of being the son of a celebrated
> foreigner, whose name is registered on the planetary system, but of
> having been the stipendiary Secretary of the Royal Society – all
> that is requisite for the honour of that elevated office.[83]

Granville believed that the Society required a President who should be
above party, and therefore necessarily of high rank, as did many. Others
identified Herschel with his supporters, especially the mathematicians and
astronomers of the Astronomical Society (though South perversely ended
by favouring Sussex, or seeming to do so). Then again, the Herschel
supporters were not very skilful campaigners. They were late in the field,
the famous public advertisement appearing only three days before the
Anniversary Meeting, which gave little time for non-London Fellows to
assemble, for after all elections were normally *pro forma* affairs. Even more
ineptly, they were overconfident, and told some Herschel supporters not
to come to London to vote.[84] Above all, although they were able to gather
support from many scientists, they could not convince them all; they had
not mounted an effective enough campaign to convince the doubters or
arouse the indifferent; while in the end, not all those who signed in favour
of Herschel's candidacy voted for him. Perhaps by his apparent indifference
Herschel showed that he was indeed not anxious for the post, which Sussex
clearly was, and an uninterested or temporary President was hardly what
the Society wanted, or indeed, what it needed. Although the geologist

Murchison (Herschel's exact contemporary, F.R.S. 1826) thought the end of science had come – he wrote excitedly to Herschel

> The *annus mirabilis* of the R.S. has arrived and we have *lived* to see it! – believe me I never in my life excited myself with such sincere devotion to a cause, as in my humble efforts to secure your election, which had science been heard was secured by a majority of *20* to *1*! The analysis of the votes ought to pass thro' Europe, to redeem the characters of the *scientific* men of the Society, for all the courtly and servile circumstances considered, they stood fearlessly to their guns[85]

– and *The Lancet* could say 'The Society is rotten to the core' while congratulating numerous medical men for supporting Herschel 'openly' (in fact the list given was incorrect, as a correspondent promptly pointed out), a good many 'scientific Fellows' had in the end voted against Herschel, and it was by no means only the bishops, clergy, army and navy, those 'supporters of the crown', who voted for Sussex.[86] The issue was a complex one. Had Herschel been more obviously a truly distinguished scientist, or older and wiser, or a more effective man of affairs, or more eager for success, he would probably have won. Whether in that case the Society would have survived intact is another question.

3

Reform and revision (1830–1848)

To those who had supported Herschel, his defeat was a sign of deepest reaction, and they saw the Royal Society as plunged back into the dark ages of the beginning of the century. Those who had supported the Duke of Sussex were by no means so unanimous, although many must have hoped that the Society would resume the aspect it had possessed under Banks. Virtually no one seems to have appreciated the fact that Sussex meant what he said in accepting the nomination, and that he intended to try by all his efforts to improve the state of the Society, raise its reputation, and drag it forward into the age of reform. Paradoxically it was the apparent victory of the forces of conservatism that was to produce, slowly, the reform the radicals wanted. It was, inevitably, slowly because a necessary preliminary (as the radicals did not perceive) was the creation of a more business-like and efficient administration which should ensure the smooth running of the Society. As it was to turn out, at Sussex's resignation in 1838 the Society was far more 'modern' than it was in 1830, even though it had to wait another nine years before a thorough revision of the statutes was at last achieved.

The Presidency of H.R.H. the Duke of Sussex

The gloomy views of the defeated Fellows must have been reinforced by the fact that once the election had been determined it was realised that the other officers and the Council must seek permission to wait on their President – for this was precisely what they had predicted, that he was too high above the Fellows for proper business communication. But in fact over the years the new President seldom enforced protocol, and was nearly always accessible at Kensington Palace when in London. He began as he meant to go on, by presiding over the first Council meeting and the first regular meeting of the Society (both on 9 December 1830), and almost immediately showed that he intended extensive reorganisation.[1] In this he was to be ably assisted by his officers. The Treasurer was not after all Gilbert, but J. W. Lubbock (F.R.S. 1829); a young man, son of a wealthy banker, a Cambridge mathematician whose special interest was in the problem of tides, he nicely combined financial acumen and scientific ability, and was to prove a zealous and efficient officer. The Secretaries were Roget and Children. Gilbert was in fact to have little influence, for the President's chief adviser was Peacock, a contemporary of Babbage, Whewell and Herschel at Cambridge and ironically a future biographer of Young, like his friends an able mathematician. The Foreign Secretary was Charles Konig (F.R.S. 1810), born and educated in Germany, but since 1813 in the natural history department of the British Museum, an able if unoriginal mineralogist. This was a group which would certainly now be considered a strong team.

None seemed to find the President's exalted rank any great difficulty, even in these early years, for he was no autocrat. As he put it at his first Anniversary Meeting (30 November 1831)

> The ostensible duties...of your President, are chiefly ministerial: he is your organ to ask and to receive your decisions upon the various questions which are submitted to you; he is your public voice to announce them,

while within the Society, he believed, there should be only one voice, not that of an autocrat but that of unanimity, for he wanted 'free expression and interchange of opinions'. And by the next meeting he was thanking the Vice-Presidents, officers and Council for their industry in the running of the Society, while congratulating the Council on its tendency to reform. At the same time he tried to increase the morale of the Fellows, declaring that

> I believe the scientific character of this country to be most intimately associated with the scientific character and estimation of the Royal Society

and, while recognising that a man of science without an independent fortune 'must pursue it, as is most generally the case, in connexion with a laborious profession', he insisted upon the strong spirit for science displayed in the country.

During these first two years of his Presidency Sussex thus acted to restore the spirit of the Society, lost in the dissensions of 1830, very much in the liberal line which he had always favoured. (Rumour declared that Peacock wrote the addresses; however true this was, and it is not at all unlikely, the general tone of the addresses conformed perfectly to Sussex's known views.) While the disaster predicted by his opponents at the time of his election failed to materialise, and while the Society's prestige certainly improved, there were admittedly to be serious disadvantages to his eight-year term of office. The first arose from the state of his health: it was not good, and by November 1834 his sight had become so severely affected by cataracts that he was unable to preside at the Anniversary Meeting, and sent a letter to say that although he was hopeful, and would accept re-election, he should not wish it in November 1835 if his sight was not improved by the operation he awaited. At that time he was forced to tell his officers 'If [the Council] wish me to remain I shall certainly not retire, and give myself another year's trial', but if then he had not recovered his sight he would 'as an honest man' retire, feeling 'bound to make the Tender, however painful the same prove to my Vanity and Pride'.[2] By November 1836 he had been operated upon and his sight restored, and so he was naturally re-elected, since he wished it, not finally resigning for another two years, although during this time he rarely presided over the Council. But although seldom at Somerset House he was an active administrator, keeping in touch through Children and Peacock, and through them furthering the reform he favoured. He had decided views on the intimate affairs of the Society: thus he always drew up a list of proposed Council members, and occasionally 'insisted' upon certain names,[3] but at the same time could generally honestly say 'I do not wish, ever, to dictate to the Society any step connected with their Interests'.[4] He supported his officers sedulously, and while insisting on his right to nominate Vice-Presidents to preside at specific meetings, in fact Lubbock took the Chair almost invariably at Council meetings while he was Treasurer (1830–35), and Baily, his replacement, did the same until he in turn resigned in 1838.

Only at times was their President's exalted status apparent, except for his continued absence from meetings, for protocol was not severe. It was time-consuming, certainly, that Children had so often to go to Kensington Palace for consultations. And it was his royal status, not his loss of sight, which was in the end the occasion for his resignation; as he told the Council by letter, he could no longer live in London nor fill the office

either with credit to myself (being unable to do the honours of the situation in a manner suited to my rank and the station I occupy in the country) or with the advantage to the character and dignity of the establishment itself the reason being, as he did not say, that the Government had refused to accept the motion put forward by Gilbert in the House of Commons for an increase in his annual grant (never as large as that of his brothers) on account of the expenses arising from the need to provide entertainment for the Fellows and foreign visitors.[5]

A curious disadvantage of his position in society was his occasional misunderstanding of middle class values and ethics – for after all, the officers and active members of the Society were largely professional men. The most dramatic and (to Children especially) troublesome instance occurred early in Sussex's Presidency over the blackballing of a candidate: kind friends told the candidate that he had been rejected firstly because put up by Pettigrew (then unpopular), and secondly because his character had been attacked at the Council meetings, an attack whose nature darkened with investigation.[6] It soon appeared that His Royal Highness had told Pettigrew (in confidence) that the candidate 'was suspected of odious practices', the source being given as Children. Now when the name had been raised at the Council, some members had expressed disapproval, and Children, appealed to, said there were '*awkward reports* afloat' (apparently of financial dealings of a dubious nature, and certainly the other Council members took the words in such a sense). To Sussex, 'awkward reports' meant moral delinquency, and to Pettigrew confidentiality was not as important as friendship or perhaps mischief-making. (He soon lost his connection with Sussex as a result.) It was all settled in the end, but it was a strain for Children and did the Society's reputation no good, while it went contrary to Sussex's fervently held wish that harmony should prevail within the Society and especially the Council, a harmony he strove to promote and which he expressed in earnestly advising the Council upon his resignation to 'avoid all matters which are of a tendency to create angry feelings or heart-burnings, on questions of religious or a political nature', absolutely in the spirit of the founders of the proto-Royal Society in the dark days of 1645. As Sussex wisely said 'They have nothing to do with science, except to create difficulties and to impede philosophical researches'.

On the whole he had been successful in preserving harmony during his Presidency, and on the whole the Society had experienced a considerable increase in efficiency as Sussex wished, and a modest amount of desirable reform. There had also been a weakening of Presidential autocracy, deliberately by Sussex himself in the early years, unconsciously through his absence in the later years. As efficiency increased, as the officers did more because they had more regular duties, and the day to day business of the Society became more open, so inevitably Council authority increased and the Fellows at least knew more of what the Council was about, even if their direct influence was no greater. (At the same time they were more deeply committed, for one of the first acts at the first Council meeting over which Sussex presided was to take steps to gather in arrears from delinquent Fellows – this not only improved the Society's finances, but meant that it became apparent that the honour of election was matched by the responsibility of financial support.)

The new President was rightly anxious to improve the quality of the ordinary meetings. As early as 16 December 1830 it was resolved that the abstracts of papers read at the meetings and entered on the minutes should be printed for 'the use of Fellows', the Treasurer and Secretaries to constitute a committee to this end. By 10 May 1832 a more ambitious scheme was decided on: abstracts of all papers printed in *Phil. Trans.* from 1830 to 1832 were to be printed together (Lubbock took charge of this and Children told him that 'science is much indebted to you' as a consequence)[7] while further proceedings (that is, brief accounts of meetings) were to be published annually, to contain both abstracts of papers to appear later in *Phil. Trans.* (and, tacitly, of most other papers read to the Society) and brief reports of business conducted before the reading of the papers. From now on the *Proceedings* (*Procs. R.S.*) became an important contribution to the Fellows' involvement with the Society's affairs, and in it were printed not only the President's address at the Anniversary Meeting (his permission always being first sought) but a report by the Council and senior Secretary, a list of Fellows deceased (with eulogies of the more important) and medal awards. Another innovation was the printing of the Council Minutes 'For the use of the Members of the Council only', beginning with those for 1 December 1832; a resolution to do so was taken on 14 March 1833, the minutes to be sent out with each summons. This meant not only that there was a readily available permanent record of the Council's decisions, but that this was now in the hands of *all* Council members, not only of those attending at Somerset House, thus obviating the need to have the bulk of the Council members either permanently resident in London or having frequent business there. (It was not until the 1850s that, with the growth of the railways, it became possible to have a non-resident Secretary.)

The President also thought a good deal about how to raise the standards of the papers read, in order to make the meetings of wider interest: he suggested that this might be achieved by his plan for reorganising the Council under six Vice-Presidents chosen to represent the widest possible range of scientific subjects, who should actively solicit suitable papers for presentation at the meetings over which they presided (presumably on a regular basis).[8] Nothing came of this particular plan, but in January 1833 the Secretaries were instructed that whenever they

> entertain a doubt of the propriety of reading any Papers which they
> receive for that purpose, they shall be authorized to refer to such
> two Members of the Council as they consider most conversant with
> the subject of the Paper in question, to determine whether it shall
> be read or not,

an obvious extension of the earlier directions to the Secretaries and a replacement for the Sub-committee on Papers, which soon afterwards ceased to exist. This admittedly somewhat informal refereeing system was taken seriously by the Secretaries; coupled with the decision that no paper might be printed in the *Philosophical Transactions* until approved in writing by two members of the Council, it necessarily raised standards. But there continued to be appeals by rejected authors, occasionally direct to the Council, and the principle (which seems to the twentieth century quite normal) of anonymity in refereeing was difficult to uphold.[9]

In 1838 a real storm blew up, which, at least in part, led to a revision of the system. It concerned Dr Marshall Hall (F.R.S. 1832), now remembered for his work on the physiology of reflex action, an irascible man who proceeded to substantiate the feeling of the physical scientists that medical men imported the disputatious character of their profession into the Society. Hall took exception on behalf of a colleague to the claims for independence in research made by an entomologist, George Newport (F.R.S. 1846), and Hall's letter to the Council provoked a heated reply. Both letters were read at regular meetings of the Society. Evidently it was felt that Roget (as the biological Secretary – although not yet so called) needed support, and on 8 March 1838 the Council resolved

> That it is desirable that a Committee should be formed, from those
> members of the Society who have more particularly attended to the
> science of Physiology, to whom papers on that subject should be
> referred, and who might also promote such subjects of
> experimental research as may be suggested by those papers,

while two months later it resolved to have a 'permanent Committee in each department of science', namely astronomy, chemistry, geology and mineralogy, mathematics, physics and 'Physiology, including the Natural History of organized beings'. Soon these Sectional Committees were entrusted with making recommendations for the Royal and Copley Medals, as well as with the refereeing of papers submitted, all of which rendered them very powerful. This 'modern' system did not, as will be evident, work very smoothly in the mid-nineteenth century, and the Physiological Committee in particular was to appear so biassed that the system broke down ten years later. That this was so illustrates the difficulty with which the established modern view of refereeing came into being, though no doubt the author of many a rejected paper in the last century and a half has doubted whether the referees were as disinterested as they were claimed to be. But that the establishment of the Sectional Committees was a step towards that smooth and representational running of the Society at which Sussex aimed cannot be doubted.

At the same time that the conduct of ordinary meetings was under consideration, Fellows from time to time demanded the right of discussion, either of papers or of the general running of the Society. The problem became acute in 1836, partly as a result of a storm arising from A. B. Granville's presentation to the Society of a second edition of his critical book, now entitled *The Royal Society in the XIXth Century*, containing a detailed analysis of the Fellows' publication record, the question being whether he should be thanked or not. After a Council ruling (28 January) that according to the statutes the business of the regular meetings was indeed purely 'philosophical', steps were taken to call a Special General Meeting (23 June).[10] At the Anniversary Meeting the Presidential address touched on the question of the introduction of discussion at meetings which, he said, had been 'raised upon the Minutes of Proceedings'; he held to the principle of prohibition on the grounds that debate led to the loss of personal dignity, as confirmed by the 'very serious amount of irritation... produced...in the course of last year, though originating from the most trivial causes'. The Council ruling about Special General Meetings was, he held, a good one, and he reiterated his opinion that debate of papers would give rise to irregularities and personalities. Hence any introduction of debate (or rather re-introduction, because the meetings were mainly discussions well into the eighteenth century) was postponed.

At the Anniversary Meeting of 1831 the Council Report (itself a novelty) could tell the Fellows of a modest degree of statute reform. This had been undertaken at the Council meeting of 16 December 1830 with the formation of a committee, to be composed of the Council with the power to add others as it saw fit. An attempt was made to enlist former critics of the Society, like Babbage and supporters of Herschel, but with little success, for most refused to serve, more or less politely according to temperament. (Francis Beaufort even tried to get an assurance that the terms of reference would include the decisions of the 1827 committee before accepting, although he did ultimately agree to serve.)[11] In the end the proposed changes were modest, far too modest to satisfy the critics. This was partly because radical changes would have meant the expense and effort of a new charter,[12] partly because radical revision meant limitation of membership and many honestly believed that only an expanding Society could be a flourishing one, especially as past Treasurers had spent, rather than invested, composition fees, so that the Society's income depended on new members. But there were some genuine changes: candidates were to be scrutinised more closely, the minimum number of signatures being raised from three to six (there were in fact few candidates who did not meet this criterion); elections were to be held four times a year only, in January, February, April and June, to avoid constant interruptions to papers (but this was found to take too much time at the specified meetings, so the limitation was abolished in 1835); the annual subscription was raised to £4 a year, but neither a bond (to ensure payment) nor a composition fee was required; the 'house-list' of Council and officers was to be printed and circulated to all Fellows a month before the Anniversary Meeting, as was the (printed) Treasurer's report; and, as noted above, provision was made for the summoning of Special General Meetings. It was not a great deal, but it manifestly improved the efficiency of the Society.

At the same time, attention was devoted to the problem of medals and lectures. Once again there was a committee, which began by instructing the Assistant Secretary, James Hudson, to prepare a printed list of previous awards, and which quickly proposed major revisions. By November 1831 it had been decided that the Copley Medal should be awarded annually to any living author of a work published or communicated to the Society, members of Council being excluded. The Bakerian Lecture was to be given at the first meeting of the new session (that is, after the Anniversary Meeting), and in 1836 careful planning for it was instituted. The Croonian and Fairchild Lectures, both of which had become monopolies, as Babbage complained, were suspended for twenty years, in spite of the death in 1832 of that inveterate Croonian Lecturer, Sir Everard Home. No change was then made in the regulations for the Royal Medals; although William IV agreed to continue them, none was awarded until 1833, when the President could proudly announce that real medals themselves were now forthcoming. At the same time new regulations were prescribed, and announced in advance, involving a three-year cycle of subjects, a regulation which remained in force until 1850 with occasional, generally unsuccessful attempts to prescribe topics as in the French prize questions of the Académie des Sciences (even when the Sectional Committees proposed such prize questions, the usual formula for the Council's resolution was, after stating the question, to add the words 'or for the most important paper in' the general subject in question submitted for publication in the *Transactions* in the previous three years). After 1850 one medal was awarded to 'each of the two great divisions of Natural Knowledge' in each year.

A very important although not immediately obvious reform was a tightening of the Council's control over financial matters. Not only were the accounts audited annually but now there was a Treasurer's report at the Anniversary Meeting, which was printed for distribution to all Fellows. All major payments, including salaries and wages, were to be submitted by the Treasurer to the Council for approval. Further attempts were made to economise where necessary, particularly in the previously somewhat lavish style of printing the *Philosophical Transactions* (then as now illustrations were expensive). And the Society's holdings of land in various parts of the country as well as its holdings of stocks were kept under review with the aim of increasing income where safely possible. This laudable activity soon resulted in a balanced budget, and even a slight excess of income over expenditure in most years.[13]

Among other aspects of the Society's activities which underwent survey and reform was the library, which had been badly neglected in the early years of the century and still wanted much attention and improvement. In February 1831 the Council decided to ask Henry Ellis (F.R.S. 1811), principal librarian of the British Museum, to supervise the arrangement of the library. A preliminary catalogue was completed by July 1831 and in November new regulations were devised, with the Treasurer and Secretaries to act as a management committee. Then in February 1832 a more ambitious project was decided upon, namely a printed classed catalogue of all the books in the library. As usual, a committee was set up for the purpose, consisting of the Library Committee with the addition of (initially) four Fellows (others were also appointed from time to time): the committee as constituted sensibly included H. H. Baber (F.R.S. 1816), Keeper of Printed Books at the British Museum. Probably it was his recommendation which led to the appointment as cataloguer of Anthony Panizzi, recently appointed as assistant librarian at the British Museum, and Baber's subordinate. Panizzi was an Italian refugee from Modena, who had practised as a lawyer before being forced to flee his country for revolutionary activities; in England he was particularly protected by Lord Brougham, who secured him the professorship of Italian at London University and his post at the British Museum, where, in 1837, he succeeded Baber and began to frame the rules for their alphabetical catalogue. (He was to become chief librarian in 1856.)

He must have seemed ideally suited to the task, and at first was quite content, although appalled to find old books in such a bad state of repair that many were useless, especially the non-scientific ones (which he estimated in 1833 to be a third of the whole). By the end of 1833 a series of classes (ultimately eighteen in all) had been determined upon and the Committee had given Panizzi assistance on where to put 'difficult' subjects, ranging from acoustics to alchemy. Panizzi certainly worked under some difficulties, for the books were in disarray, and although he was given the assistance of J. D. Roberton (librarian assistant to Hudson, the Assistant Secretary) it was not an easy task. The physical difficulties were increased when Panizzi so clearly pointed out the deficiencies in the matter of storage and shelving that work was put in hand to provide more shelf space in a specially built gallery; shortly thereafter the President secured additional rooms to house the library.

Still, things went reasonably smoothly until, in 1835, Panizzi sent the first slips to the printer. He then told the Committee that he alone was responsible for corrections and revises and that he had told the printer that no one else should interfere; inevitably the Committee was distressed and insisted that all proofs should be seen by at least some of its members. At this point dispute and debate set in. Panizzi was a perfectionist (he needed, he said, four revises because he could not correct everything at once) and his idea of a catalogue was one complete in every bibliographical detail. The Committee members were book users, and wanted simplicity both for ease of reference and for economy. The Committee in 1836 offended Panizzi deeply by resolving that 'All notes expressing matters of opinion on the articles in the Catalogue be omitted', with no use of the first person, and by insisting on the right of the Committee members to supervise the proofs closely. No doubt Panizzi was right to find this tedious and delaying, especially as he felt it to be 'his' catalogue. He not unreasonably saw himself as a professional and knew that as a cataloguer he was striking out new ground; but also he was touchy, and inclined to feel that he was unfairly treated as a foreigner. The President in his 1836 Anniversary address had spoken of the catalogue as being composed and revised by members of the Council; Panizzi replied in a printed pamphlet 'defending' himself, and attacking the Council for not paying him all the monies due. To the President's Anniversary address in 1837 was appended the Council's justification of the Committee, an insistence that Panizzi had been fairly paid, and a statement of the severing of his connection on 14 July 1837. In January 1838 a member of the Catalogue Committee defended the omission of Panizzi's bibliographical notes in a letter to which Panizzi in turn replied in a printed letter addressed to one of the members of the Committee. In December, the Council having declared the matter closed, Panizzi began a long correspondence about money still due to him, and throughout 1839 prolonged negotiations took place over submitting the matter to arbitration, negotiations handled tactfully and patiently by Northampton, now President: Panizzi received a good part of what he claimed due, and the catalogue (shorn of any trace of Panizzi) duly appeared in the same year. It had been an unpleasant interlude, not creditable to anyone, but it had produced order in the Society's library, and a very good straightforward catalogue of its holdings.[14]

Several important changes of office occurred during the later years of Sussex's Presidency, changes which did not greatly alter the general running of the Society. Lubbock resigned as Treasurer in 1835, to be replaced by Baily, who now acted as a senior Vice-President, presiding over most of the Council meetings. Lubbock's resignation seems to have been partly occasioned by the rumour that he had acted in such an arbitrary and imperious manner as to occasion the resignation as Assistant Secretary of James Hudson.[15] Hudson was replaced by his assistant, J. D. Roberton, in spite of the fact that Hudson had complained of Roberton's uncooperative spirit; this suggests that Hudson himself was not easy to work with, while Roberton had managed to work with Panizzi, and seems to have been quietly and unremarkably hard-working.

It is notable that Baily, formerly among the leaders of the astronomical rebels, apparently never tried to institute reforms during the years when the Royal Society had an absentee President. Lubbock was to replace him again in 1838. In 1837 Children retired as Secretary, pleading ill health and pressure of work, to be replaced by S. H. Christie (F.R.S. 1826). Christie, a successful Cambridge mathematician who had made significant scientific contributions to magnetic theory, was currently a hard-working and reforming professor of mathematics at Woolwich, but although he served as Secretary for seventeen years, retiring only at the age of seventy in 1854, he seems to have made little impact on the Society's affairs, leaving Roget, five years his senior, to act as senior Secretary. At the end of Sussex's Presidency the officers were all strictly scientific men of some repute.

A second attempt: Northampton's Presidency 1838–48

When on 14 September 1838 the Duke of Sussex's letter of resignation was read to the Council, only six members were present. It is thus worth noting that one of these was the Marquis of Northampton (F.R.S. 1830) who

attended for the first time since his election to the Council in the previous
November. He also attended the next Council meeting on 1 November
(along with thirteen others) when after the defeat of a resolution to limit
the President's tenure to five years, the Council voted to recommend
Northampton as the next President of the Royal Society. His qualifications
were, it might seem, somewhat curious, and indeed it is clear that he was
not universally approved: a disgruntled Fellow and sometime Council
member wrote in 1845 that the Council, wishing to descend from royalty
to the nobility had a choice of the Earl of Burlington (F.R.S. 1829, Council
member 1836–38) or the Marquis of Northampton, and so, 'of course' chose
the higher title and the lesser scientist.[16] Both Northampton and Burlington
(William Cavendish, later Duke of Devonshire, the future donator of the
Cavendish Laboratory to Cambridge) had been educated at Trinity College,
Cambridge, eighteen years apart, the younger, Cavendish, having taken a
distinguished degree. Northampton, although not personally impressive,
had an active interest in geology, possessed a notable collection of minerals
and fossils, and had shown himself an exceptionally able patron of science
throughout the 1830s in his attendance at the British Association for the
Advancement of Science. Burlington had never been a whole-hearted
Fellow and, although he had attended Council meetings frequently, he was
now abroad; besides his presidency of the British Association in 1837 had
been reluctant and undistinguished.

The 1830 rebels, Murchison particularly, at last saw that Herschel, just
back from his astronomical sojourn in South Africa, was not a practicable
or even desirable candidate, and with Whewell and Peacock (who were on
the Council, as he was not) moved to support Northampton. That other
scientists thought him a suitable candidate is confirmed by the opinion of
so distinguished a mathematician and astronomer as William Rowan
Hamilton (rather curiously never F.R.S. although he had received a Royal
Medal for a paper published in the *Philosophical Transactions* in 1834 and
was to be president of the Royal Irish Academy in 1837 and Royal
Astronomer of Ireland; he attended some meetings of the British Association
in England, while never serving as president). He wrote to Lubbock on
5 November 1838

> I gather, from the Athenaeum of this morning, that you are likely
> to resume your connexion with the Royal Society. It gratifies me
> also that, since it was understood that Herschel would not serve,
> the Council have decided on recommending Lord Northampton to
> succeed the Duke as President. I spent the greater part of last

Fig. 8. Spencer Joshua Alwyne Compton, 2nd Marquis of
Northampton (1790–1851), F.R.S. 1830, P.R.S. 1838–48.
By Thomas Phillips

Fig. 9. J. G. Children (1777–1852), F.R.S. 1807,
Secretary 1826–27, 1830–37.
By (?) S. Pearce

month with him (Lord N.) at Castle Ashby & in Liverpool, & very much enjoyed the opportunity of improving my acquaintance with one whom I esteem & love so much.

It would seem, then, that Northampton had wide support from all sections of the Fellows, and rightly so.[17] For he represented that spirit of reconciliation so warmly advocated by Sussex and of which the Society was still in need. He also had, clearly, far better claims than Sussex for the position of P.R.S. in terms of scientific attainments; although not a distinguished scientist like later Presidents, he had scientific knowledge and interests of a serious kind.

In this spirit he set to work conscientiously and with considerable attention to detail.[18] As already noted, he dealt with Panizzi faithfully, patiently, sensibly and without malice (he later consulted both Lubbock and Children as to the advisability of including Panizzi in his general invitation to the officers of the British Museum to attend his soirées). He evidently answered letters promptly, even though, chiefly in his first year, he often left the onus of presiding over the Council to Lubbock, Treasurer once again 1838–45. (Lubbock was to be succeeded by George Rennie, F.R.S. 1822, a civil engineer like his more famous father and brother.) He found the Royal Society's autumnal meetings inconvenient, although, except in 1841 when he was abroad, he dutifully attended the Anniversary Meetings. In 1845 he proposed an innovation, which he only put through in 1848: in this, his last year as President, a 'General Meeting for the election of Fellows' was held early in June and he then gave his Presidential address, although there was still an Anniversary Meeting; his successor was only to retain this new custom for one year. It was indeed more convenient for Court and Parliamentary circles than for academic or professional ones; and the St Andrew's Day Anniversary was too well-entrenched a custom to be easily given up.

As President, Northampton took his social responsibilities seriously. He gave several soirées every year, usually with a dinner to the Council beforehand. He also provided tea and coffee in the Library after the regular meetings of the Society (now at 8.30 p.m. and still lasting about an hour). These informal social occasions were warmly welcomed and became a desirable feature of Society life, so much so that when he resigned Northampton declared that he thought the Society, not the President, should pay for the refreshments served after the ordinary meetings.[19] He also saw (apparently being the first to do so) that the soirées should be held in the Society's rooms, even though this would mean some remodelling, for again this would make them more peculiarly the Society's own. This is the first suggestion for the modern conversaziones. When he was in town Northampton always, other duties permitting, attended Council meetings, Club dinners and regular meetings; he was also active in the Geological Society and Club, and was faithful in attendance at the House of Lords. He did not ever neglect correspondence, although he ruefully remarked to Lubbock in 1839 'I am afraid the council may think that they have got a very crotchetty President & one who is too busy by half'. He was indeed busy, and his attention to Royal Society affairs shows that he took all business seriously.

But he was not by nature an innovator, nor, apparently, did he see the need for instituting reform, although he does not seem to have opposed it when others sought it. When reform did come in 1847 it was to be even more wide-ranging than the proposed reforms of the Charter Committee of twenty years before, going through in the wake of such a great public quarrel that opposition seems to have been quite swamped in the public outcry. Yet, ironically, in a quiet way steps had already been taken about the handling of papers that should have prevented the scandal of 1845. For (on 14 March 1839, with Northampton in the Chair) Daniell proposed to the Council sweeping changes in the handling of papers, changes approved at the next Council meeting (11 April). It was decided that abstracts of all papers read to the Society should be 'forthwith' printed, and copies sent to all members of the Council and of appropriate Scientific Committees within the reach of the twopenny post (this became to all members with the introduction of penny postage in 1840), accompanied by a 'circular letter' stating that the original papers were available for inspection in the Society's rooms and requesting Council members to let the appropriate Committee have their opinions as to whether the papers should be printed or not; further, all Committees were empowered to send papers to referees, whose reports were to be transmitted to the Council (which, it must be remembered, also acted as the Committee of Papers). It had already been emphasised that the Committees should meet monthly. This efficient and modern-sounding procedure should have improved standards and made acceptance or rejection of papers more satisfactory. Although it apparently worked well for five years, it was ultimately to collapse because it could not cope with a crisis of confidence in the impartiality of some Committee members.

At the same time the procedure for electing candidates was being scrutinised, and in May 1839 the Council constituted itself a committee to draw up new forms for certificates. In January 1840 the Council, having been made aware of the state of the Donation Fund and the slight use made of it in the dozen years since Wollaston's death, instructed the Scientific Committees to report to the Council whenever they came across suitable 'objects...for the disposal of any portions of the Donation Fund', a beginning of its serious use.[20] Slowly too the Council began to show stirrings of an interest in reform, although the minutes are curiously unenlightening about details: for example on 15 October 1840 when it was resolved to print 500 copies of the statutes 'corrected to the present time', the reason for this is not stated. But the desire for reform was there, and was expressed from time to time.

Serious, although at first quiet, moves began in 1845, when on 5 June it was resolved without fuss that discussion should be allowed during the next session 'under the direction of the President or Vice-President in the Chair', with the Council retaining the right to make regulations for it. With this went the practice of announcing at the end of each meeting the subjects of the papers to be read at the next meeting. There is a sense in which this practice may be said to reflect a somewhat belated recognition of the obvious fact that no one could by this date be equally interested in all branches of science. The President, in referring to it in his address on 1 December 1845, called it undoubtedly useful, but he was still sceptical of the benefits of discussion, which in his opinion certainly required control by the presiding officer. Some idea of what occurred at the early meetings where discussion was allowed (it seems to have been a little slow to develop) may be gleaned from the report (by Walter White) of a lively geological evening on 23 March 1848:[21]

> Dr. Mantell's paper read this evening at the Royal Society supplies some omissions in Professor Owen's paper on the same subject, for which the Royal Medal was awarded some few years since [1846 'A Description of certain Belemnites']. A warm discussion ensued; first Mr. Christie grew angry because Dr. Mantell wished to read a short supplement. Mr. Owen spoke at some length, throwing discredit and contempt on the whole paper. Mr. Bowerbank in favour. Then Dr. Buckland in a most luminous and humorous discourse, then Dr. Carpenter inclining to Mr. Owen's view. Mr. Gray made a few remarks, and at past eleven o'clock Dr. Mantell replied. The most interesting meeting which I have yet known at the R.S. Mr. Owen seems unnecessarily severe. He however does not like to have his views contested.

This meeting was clearly a great and desirable change from the dull reading of a paper of thirty years before, but its very liveliness verging on acrimony was evidently what Northampton dreaded. Something about the Royal Society atmosphere seems to have bred suspicion, mistrust and impatience; as White had remarked earlier in the year (9 February) 'self and not science is the prime mover', instancing Sabine, De Morgan and Brewster, and adding 'When will philosophers be true to their calling?' Harmony was slowly to become more usual as the century wore on, and manners grew milder.

Certain administrative reforms had taken place, in the normal course of events, and these had nearly all been beneficial to the smooth running of the Society. When in November 1843 the Assistant Secretary, Roberton, died, it was decided to make the post full time and resident by combining with it the duties of the librarian, and C. R. Weld, the Society's future historian, was appointed, serving until 1861. It seemed that he in turn required an assistant and in May 1844 Walter White was appointed, to be variously called attendant and sub-librarian; he was to succeed to the Assistant Secretaryship in 1861. After resigning in 1885 White left a journal later published by his son, which contains much revealing information. The previous librarian, Shuckard, who had done valuable work in locating, assembling and cataloguing manuscripts, had unfortunately gone bankrupt earlier in 1843, an action then regarded as though it were a moral character defect.[22]

Then in 1846 it was rather suddenly decided to undertake more serious reform, the Council Minutes for 7 May recording without explanation the resolution 'That it is expedient to revise the Charters of the Royal Society, with a view to obtaining a Supplementary Charter from the Crown', the officers being appointed a committee 'to consider and report' on desirable changes. At the next Council meeting (28 May) the Charter Committee was directed to review the whole of the charters and statutes to see how far changes were needed, and whether a new charter was required or whether new statutes would suffice. Significantly W. R. Grove (F.R.S. 1840) was then added to the Committee which now included Leonard Horner (at this time president of the Geological Society). Grove already had a considerable reputation for his work in electricity and electro-chemistry as well as having practised as a barrister (he was later to be eminent in the legal profession), and had been elected to the Council in 1845. Clearly he was well suited to work on the Committee, and was later said to have played a significant part in pushing through reforms, particularly the limitation of membership numbers;[23] it is not without significance that the minute following his appointment records the Council's decision to discuss whether it would be 'expedient' to limit the number of Fellows elected each year in any revised charter. The Committee met in time to report very fully to the Council on 18 June with detailed recommendations on the election of candidates, proposing a limitation of fifteen in the number of new Fellows to be elected each year, these to be recommended by the Council for the Fellows to vote upon; the report added a detailed statement of the financial consequences which would arise from such a limitation.

Over the recess the Committee was to seek legal advice as to whether the proposed changes required a new charter. The legal opinion then was that it was not possible specifically to limit the numbers elected without a change of charter, which would be expensive and troublesome. This difficulty was overcome by changes in the statutes of a less rigorous kind. It was perfectly legal to pass a statute by which the Council was to present a list of only fifteen candidates annually, sent to the Fellows a fortnight before election day, which was to be on one day of the year only (at this time the third Thursday in June), and to revise in minor ways the information demanded on candidates' certificates.

All this the President was to announce at the Anniversary Meeting, stressing that the intention was to secure higher standards. But, as one Council member was to point out, what mattered was less the number of candidates than their quality; he wrote[24] that he was anxious that the Council should regard itself as pledged to elect only a man known to be either

> A *distinguished* cultivator of Natural Science, or eminent as a patron
> of science; he believed that if proposed candidates were so
> denominated the problem of limitation would not arise, and hence
> the Council would neither omit good men for want of room, nor
> elect inferior men to make up the numbers.

But most saw limitation as the best way of securing the higher scientific standards so earnestly sought by the reformers. However, finances were seen initially by the Council as a great problem: a review by a Finance Committee under Horner, a successful man of business as well as an eminent geologist, resolved the difficulties and in mid-January 1847 the Council asked Horner, Grove and the officers to draw up new statutes to embody the suggested reforms.

These, voted into law on 10 February 1847 after due consideration (the Council was meeting weekly, so anxious were all for action) were clear and forceful. As indicated above, the number of Fellows belonging to the Society was to be limited by the expedient of the Council's presenting a house-list of only fifteen candidates, chosen as the most eligible (as indicated by their certificates) of those whose names had been presented by the Fellows at large, it being understood that eligibility was to mean distinguished by contributions to or promotion of natural science, and election was to take place once a year only. The Council was to be elected separately at the Anniversary Meeting; there was to be a house-list consisting of eleven members of the existing Council and ten Fellows not members of the existing Council, while the officers, also recommended by the Council, were to be chosen from this list. The dominance of the Council in future affairs was clear; on the other hand its activities were to be manifest and open to scrutiny, so that it was unlikely that it would be tempted to diverge from the principles which lay behind the proposed limitation of Fellows. The Privileged Class of Fellows (princes, peers and privy councillors) remained, to be elected whenever proposed, their election not to be assimilated to that of the ordinary Fellows until 1874. And so, halfway through the nineteenth century, the age of the (almost) exclusively scientific Royal Society had come into being with virtually no opposition, and under a President of only modest scientific attainment.

In confirmation of this triumph and to preserve its achievements some of the most active reformers established the Philosophical Club.[25] This, like the eighteenth-century foundation, the Royal Society Club, regularly held a dinner on meeting days, and like it again, was limited in numbers: there were always to be forty-seven members to commemorate the year of its foundation, and these were all to be Fellows who had published papers in a journal of one of the chartered scientific societies. As its minutes record

> The purpose of the Club is to promote as much as possible the scientific objects of the Royal Society, to facilitate intercourse between those Fellows who are actively engaged in cultivating the various branches of Natural Science & who have contributed to its progress; to increase the attendance at the Evening Meetings & to encourage the contribution & the discussion of Papers.

Like the later *x* Club (see Chapter 4) it was a place for discussion, and many important matters were there considered, where Council members and ordinary Fellows met together, so that matters for future debate at Council meetings could be discussed informally. Though it continued in existence throughout the century its major importance had declined by the last quarter of the century as the Society became more unified and more obviously restricted to scientific Fellows.

Medals and Sectional Committees

But at the very moment when the Society was quietly and efficiently going about the business of internal reform, the officers and Council were under attack for the manner of their award of medals, and the impartiality of the Sectional Committees was being questioned. On one occasion a public scandal was created on a scale as great as those generated by the events of 1830. All this was inevitably to make the reforms of 1847 a subject of criticism. It all began in 1845, when one Royal Medal was to be awarded for physiology; ironically, that importation into the Royal Society of dissensions within the medical profession which Babbage had long ago claimed to exist was at last a reality, and *The Lancet* enjoyed the occasion to the full with detailed and vituperative report and comment over some eighteen months. The whole affair was complicated, and in the end highly discreditable to Roget, against whom *The Lancet* had waged unceasing war, partly as an entrenched officer of the Royal Society (there can be no doubt that twenty years as Secretary was too long, especially for a man of his temperament) and partly as a representative of University College London, which, in its beginnings as the University of London, he had helped to establish, and of which Wakley, *The Lancet*'s editor, fiercely disapproved.

The bare facts of the case are easy to establish, especially as they were later to be publicly reviewed, but so many side issues were imported into the affair that some detail is necessary.[26] On 27 October 1845 the Physiology Committee met after the summer recess; it rejected one paper for *Phil. Trans.* and recommended another for printing, proposed Purkinje for Foreign Member (he had to wait until 1850), and, for the third time, recommended Robert Owen for the Copley Medal. (Later (6 November) it altered the recommendation to Schwann who duly received it. This was because Owen was ineligible as a Council member – Roget who was present should have known this – and had to wait until 1851 (he received a Royal Medal in 1846).) Up to this point William Lawrence (F.R.S. 1813), surgeon, shortly to be appointed president of the College of Surgeons, was in the Chair, but being told that there was no more business, as he later declared, he 'dissolved' the meeting and left; in haste as one member remembered. After which, while the other members present (Bostock, Todd, Bowman, Kiernan, Thomas Bell and Sharpey) were talking together, Roget told them that he had made a mistake, misreading 'physiology' as 'physics', and on a closer look realised that the Committee was after all to make a recommendation for the Royal Medal this year. So the Committee was reconvened with R. B. Todd (F.R.S. 1838), physician and professor of

physiology at King's College, London, in the Chair, proceeded to consider
a list of sixteen papers, and decided to recommend a paper by Thomas
Snow Beck, a pupil of William Sharpey, which had never been referred
nor, consequently, printed. (Beck was, in spite of the subsequent publicity,
to become F.R.S. in 1851, but he had an undistinguished career; Sharpey,
F.R.S. 1839, who had lectured on physiology at University College
London since 1836, was an able though not distinguished physiologist, but
he was a notable teacher, introducing Continental ideas into England.)

   This was all very irregular, but Lawrence signed the minutes at the next
meeting, while on 30 October the Council accepted the recommendation.
Once again, this was irregular because there was no written report, and the
Council's action was subsequently rescinded and erased from the minutes.
On 20 November the Physiological Committee (which had met a fortnight
before without further action on Beck's paper) voted to ask Drs Sharpey
and Todd to draw up a report on the details of Beck's paper ('On the nerves
of the uterus'), apparently at the request of the Council, which met on the
same day:[27] at the same time the Committee, somewhat belatedly, determined
to send the paper 'to the several members of the Committee of Physiology
resident in London'. On 27 November Sharpey and Todd reported
enthusiastically in favour of Beck, acknowledging 'one important fact has,
it is true, been already pointed at in the recent published work of Todd &
Bowman, but the author of the paper has nevertheless the merit of arriving
at it independently, by his own observations' and gone further; and on this
basis the Council on 1 December duly voted the Royal Medal for Physiology
to Beck, and ordered publication of his paper in *Phil. Trans.*

   Now someone on the Committee or the Council had already 'leaked' the
proceedings, so that the actions of the Committee were known, for on
24 November T. Wharton Jones (F.R.S. 1840, since 1844 professor of
ophthalmology at University College London where he later taught Joseph
Lister; he was an active research physiologist whose own paper on blood
corpuscles had been considered for the Royal Medal) wrote to protest that
Beck's paper, to judge from its abstract, seemed to be 'descriptive of *mere
anatomical dissection*, by which no new fact is demonstrated nor any new
inference of physiological importance warranted. The subject of the paper
moreover is not new to the Royal Society' and would not improve the
Society's reputation – a dignified protest then spoiled by his claiming that
his own paper was better than Beck's. Robert Lee (F.R.S. 1830) wrote a
similar letter on 1 December, but he had some right to complain since he
had worked for many years in the same field of embryology that Beck now
entered so dramatically, although as a former member of the Council
(1842–45) Lee should have found a more effective approach than the

slightly hysterical series of complaints he now addressed to the Secretaries when the Council sent him no answer to his first letter (read at the Council meeting of 1 December 1845).[28] Lee soon saw that the best approach was to insist upon the illegality of the award, but Christie replied that Royal Medals had in the past been awarded for work not published in the *Philosophical Transactions* – an astonishingly naive admission.[29]

In March Christie proposed that the wording of the announcement of future awards of Royal Medals read no longer 'the most important paper communicated to the Royal Society for insertion in their Transactions' (a slightly ambiguous phrase) but instead read 'communicated to the Royal Society...,and printed in the *Philosophical Transactions*', a change approved by the Council on 2 April 1846. (This new wording would clearly make the award to Beck immediately illegal, but the former wording did not quite do so, since obviously he had hoped for publication, although the paper had not been approved for publication when the award was voted.) It was intended as sufficient answer to Lee's latest letter, since the Secretary was to inform him of the change, at the same time that the Council refused to 'revise the grounds on which a Royal Medal has already been awarded', not considering that it was authorised to do so. Being now busy with charter revisions, the Council presumably hoped to hear no more about the Royal Medal award of 1845.

But clearly Lee felt deeply injured, and failing to get any satisfaction from the Council, told his friends and colleagues of his grievances. *The Lancet*, being already well launched upon a campaign to expose the abuses inherent in 'the extra-professional honours attainable by medical men' was only too pleased to be able to include the Fellowship of the Royal Society in its denunciation, writing in March 1846 that 'the Royal Society, or rather, the medical section of it, has degenerated into a mere clique', attacking the fact that the office of Secretary seemed to have become perpetual and complaining that many good men were not on the Council (especially R. E. Grant, professor of comparative anatomy at University College London, F.R.S. 1836, who in *The Lancet*'s view ought to have had Sharpey's chair), and attacking the invitation to Roget by Gilbert to write a Bridgewater Treatise.[30] The editor's leader may have been stimulated by a letter printed in the same issue by one 'George Redford Esq, Surgeon, London' (not a Fellow) who upheld the rights of Lee over Beck, claiming that the former had performed many dissections, the latter only one, and wanting to know why there had not been a Committee to view their dissections and judge between them, adding 'the Council of the Royal Society is clearly ambitious of being the last Old Sarum of the scientific world'.[31]

By the next month Marshall Hall's feelings of abuse came out again, in a letter offering to relate at length to the editor *his* grievances, principally over his dispute with Newport (to whom *The Lancet* later incorrectly ascribed a Royal Medal; in fact, as Hall knew, the Physiological Committee had in 1842 recommended Hall for the Copley Medal).[32] There followed a series of letters about Roget's possible debt to Grant (which Roget stoutly denied), attacks on Roget and defence of Grant and Hall, letters by Lee and Newport (in defence of his own originality), a claim by Beck that Sharpey knew nothing of his work and a letter from Todd, while at the end of May came a review of the history of the medal regulations and various awards of which *The Lancet* was highly and indiscriminately critical. Wakley triumphantly proclaimed 'The press is all-powerful' and he intended to prove it, telling the President that 'The proceedings of the medical section for the last nineteen years are every where viewed with distrust', and that it was time that Roget retired. He added in the next issue 'The agitation is not personal nor professional, but scientific, involving the honourable conduct of the Society towards Physiological Science', proceeding to give his opinion of almost all the members, past and present, of the Physiological Committee – who were either too old or too recently Fellows, or in some other way unsuitable – adding that this was not a medical squabble, but a dispute about physiological *science*.[33] Soon, however, *The Lancet* turned to attack the College of Physicians and temporarily left the Royal Society in peace.

But in the issue of 10 October 1846 Wakley returned to the charge, with the 'startling announcement' that comprehensive changes were being 'contemplated' in the Royal Society and that Northampton, opposed to all change, had resigned. Northampton wrote to *The Times* (in a letter quoted in *The Lancet* for 24 October 1846) denying either that he had resigned or that he opposed change, but *The Lancet* continued what Northampton called its 'mistakes and misstatements' in harking back to Beck and Lee, as well as being inaccurate about the composition and activities of the Charter Committee. (It should be recalled that this committee had begun work in May, and by June had formulated its proposals, and was considering the legal aspects.)

Finally, in January 1847, fourteen months after the award of the Royal Medal to Beck, the whole affair began to come out into the open. On the 14th the Council considered a letter dated a week earlier signed by Jones and ten medical colleagues, including Lee, Hall and Grant, asking in due form for a Special General Meeting to consider the award, which the Council decided to summon for 11 February (after which the Council resolved 'That for the future no letter be inserted in the Minutes without a resolution to that effect', no doubt because of the tendency for dissident Fellows to go over the Minutes in search of irregularity). It then proceeded to consider the more urgent and important matter of finances and charter alterations. *The Lancet* crowed and congratulated itself: 'Wherever there is hollow pretence, wherever there is reward or honour unduly conferred, there is quackery, and it is the topmost bough of quackery that we have attacked'.[34] The meeting itself, reported by *The Lancet* in full detail, began almost in farce, because a question was raised about its legality, both in the wording of the original requisition and the responsibility of the action of the Council – for as Northampton pointed out, only he and the officers remained of the Council of 1 December 1845 – but the President's common sense and placatory temperament triumphantly set all legal quibbles aside, and allowed Jones to state his case, which he was able to do because he had been allowed access to both the Council Minutes and those of the Committee of Physiology. After rehearsing the events of 1845 in detail, Jones proposed a resolution to the effect that the regulations had been contravened and therefore the award of 1845 'ought to be considered null and void', a resolution seconded by a fellow signatory, Dr James Copland (F.R.S. 1833, a physician and medical writer), although he noted that now that the irregularity had been made public so that it could not occur again no resolution was really necessary. Northampton admitted that the Council had been in error, merely pointing out that everyone was confused about the exact terms of the regulations, while a member of the Committee of Physiology tried to point out various mitigating circumstances, which really only made things worse. Gray then moved an amendment to Jones's resolution, although as he said he had originally opposed the award, namely, that as the President had 'expressed...an opinion of the irregularity which attended the award and there were now new regulations, no further proceedings should be taken' – a sensible amendment eventually carried in spite of attempts to raise side issues (Babbage, for example, tried to re-introduce his old grievance about the 1836 medals).

There the affair should have ended. But Wakley could not resist crowing and attacking Roget, Christie, Bell, Newport, Todd, Sharpey and Northampton, as well as the new statutes which gave decisions on elections to the Council, not to the Fellows at large. Inevitably Roget's announcement at the 1847 election that he would only serve for one more year was greeted by *The Lancet* with great delight. Throughout 1848 Wakley continued to attack Northampton as an 'autocrat' and to report at length on Lee's correspondence with the President about a paper which Lee thought he had been unfairly compelled to withdraw, but gradually even Wakley tired of the subject.

And so the affair ended, certainly not to anyone's credit save possibly that of Northampton who had acted as an honourable man in taking much of the blame and had tried to maintain calm good sense. It had provided much gossip for the journalists and had been exceedingly unpleasant for Northampton. It was bad for Roget's reputation, not altogether unfairly. He had been a hard-working Secretary for twenty-one years by the time he resigned; although he had committed a number of acts of haste and carelessness and had, to say the least, shown lamentable want of tact, it is not clear that he felt himself at fault, for when he resigned at the end of 1848 he was almost seventy and probably did so on account of age rather than because of Wakley's attacks.

It is relevant here to consider another only slightly later case when the award of a Royal Medal created difficulties between the Council and a Sectional Committee, this time that for geology. On 16 November 1849 the Committee, with Rosse (then President) and Roget present, resolved to recommend to the Council that the Royal Medal for Geology be presented to Edward Forbes for a paper on glaciers, much to the disgust of more conventional geologists like Murchison, Buckland and Lyell (in fact main support came from Owen, chiefly apparently because he was jealous of the other serious contender, Gideon Mantell). At the Council meeting on the same day, probably at the instigation of J. P. Gassiot (F.R.S. 1840) and Horner, the recommendation was ignored, and on 22 November the Committee was directed to reconsider the award, apparently as a result of a letter of protest from Mantell. On 26 November the Committee did reconsider the matter and, with warm support from Buckland and Murchison, reversed its decision and voted to recommend the award of the medal to Mantell, a recommendation which the Council accepted.[35] The Committee displayed thus considerable irregularity, both collectively and individually: not only was Owen present at its meetings (he was not officially a member) but Murchison told Mantell much about its deliberations, as Gassiot did about those of the Council. These facts must be borne in mind when considering the subsequent abolition of the Sectional Committees.

The succession: 1848

Meanwhile, on 22 February 1848, Northampton addressed a letter to 'George Rennie, Esq or whoever is in the Chair at the present Council', in which he formally resigned (he had indicated as much verbally at the last Council meeting) and tactfully announced his intention of absenting himself from the meeting on 24 February.[36] Although he had been mildly opposed to statute changes, and all earlier motions to limit the Presidential term had already been blocked, he now remarked that ten years was perhaps too long a term, although extreme limitation would also be undesirable; perhaps eight or nine years would be best. On the whole the Council continued to dislike the idea of any statutory limitation and only after 1871 did a maximum of five years become even customary.[37] As already noted, Northampton proposed social changes, namely that the President should no longer personally dispense tea and coffee after meetings, and that, when space permitted, the soirées should be held in the Society's rooms, not in the President's house: the first change waited until 1860, but the second was effected in 1849.

The greatest innovation in connection with Northampton's resignation was the manner of it, over nine months before the Anniversary Meeting and without direction as to his possible successor. Further, although he presided over most of the subsequent Council meetings during the session he appears to have allowed full scope to all the Council suggestions except those on the limitation of Presidential terms of office. There is some significance in the emergence as active canvassers of Fellows who had apparently played little part in the Society's affairs previously. Limitation of the President's term was the particular concern of Sir Henry de la Beche (F.R.S. 1819), a geologist who would probably have been active in subsequent years had he not become seriously ill. Lyell (F.R.S. 1826 and now newly on the Council) emerged as a leading figure in negotiations for the Presidency, for with Sabine he was deputed 'at a full meeting of the Royal Society' on 9 March 1848 to approach Herschel to request him to accept nomination, a probable choice already made public by *The Lancet* five days earlier. When, as ten years earlier, Herschel refused to allow his name to go forward, various other names were canvassed: in reporting Northampton's last soirée on 22 April *The Athenaeum* mentioned the names of Herschel, Northumberland, Wrottesley, Westminster, Peel and Rosse, expressing the conviction that Rosse would be chosen. Of these, only Rosse seems to have been approached, although Northumberland (F.R.S. 1816 as Baron Prudhoe, now on the Council, a naval man and patron of learning) was seriously considered. Rosse wrote to Sabine on 23 March

> I must say that I should look upon the Presidency of the Royal Society especially under the circumstances you mention as the highest honor, and as one which would be no less gratifying to my friends than to myself,

but adding that he foresaw difficulties: he was worried about the social side of the office, since it would mean more residence in London than usual, and getting a suitable house.[38] But although he suggested Northumberland as more suitable, the Council's choice remained Rosse.

He now appears a natural choice: like Herschel he was an astronomer, and already a notable one (his discovery and study of spiral nebulae with his 72-inch reflector were to be made during his years in office), he was a rich man, active in public life, an elected Irish member of the House of Lords, president of the British Association at its Cork meeting in 1843, on the Royal Society Council 1834–35 and 1842–43. Ironically he was apparently not yet well known in general scientific circles for 'F.R.S.' in *The Athenaeum* (29 April) deplored the idea of 'another member of the aristocracy'; ironically, for Rosse's only disadvantage was that he was just what this particular F.R.S. claimed to want, an active scientist, and at the height of his powers he was reluctant to take time off from observing to fulfil his duties to the Royal Society. It is worth noting that White regarded him as 'the reformers' candidate'.

Rather more controversy arose over the filling of Roget's post as Secretary. De la Beche had intended to raise the question on 25 May, but he was absent, so Lyell announced that he would do so, and on 6 July he moved that Grove be approached. This was an obvious move, except that his election would mean having two physical Secretaries (Christie was not to retire until 1854, at the age of seventy, after seventeen years in office). Consequently Robert Brown, in response to the request of the Physiology Committee (4 July) that the Council consider 'the advantages and desire-ableness' of having one of the Secretaries 'conversant' with their subject, proposed Thomas Bell, who was, however, not chosen by the Council. There was over the next months some grumbling about the 'house-list', although *The Lancet* expressed approval, preferring Grove (a known reformer) to Bell, who was there described as 'only' a zoologist (it was forgotten that the famous Committee was properly designated as 'Zoology and Animal Physiology', but Bell was secretary of the Committee and so disliked by *The Lancet*, which also attacked Christie as an entrenched office holder). Surprisingly, at the election on 30 November 1848, Bell was elected (134 votes to Grove's 108) because somehow a 'cross list' with Bell's name was printed. The matter was discussed at the Council meeting on 25 January 1849, but nothing was done since Northampton took responsibility and apologised, and it was clear that no precedent had been created. Almost certainly Bell's election was a spontaneous demand for more official recognition of biological subjects, as evidenced by a stream of letters in *The Athenaeum* in both July and November, with the chief interest now not in the choice of President but in that of Secretary,[39] a foreshadowing of things to come.

4

How reform worked: the running of the Society 1848–1899

In the wake of the 1847 reform of the statutes came the innovation of Rosse's election, an innovation disguised as convention, since to many it seemed like a continuation of the old tradition of 'noblemen presidents'. It was not at all obvious that with the new statutes of 1847 the power of the President had been severely diminished and the power of the Council had been greatly increased. Yet this was the case. Never again could the Royal Society be seen as a monarchy, although it might on occasion be seen as an oligarchy. During the second half of the century, slowly but surely, an efficient system of working was established, in which the main authority lay in the hard-working Council (successive Councils refused to elect to their number those it thought unlikely to be faithful in attendance), with the President its chairman rather than its leader, and the Treasurers and Secretaries highly conscientious executive officers working together in general harmony for the common good. The officers, now generally *ex officio* members of all committees, gained a wide knowledge of the Society's affairs and such prestige that more commonly than not service as Secretary, Foreign Secretary or Treasurer came to be considered as a recommendation in a Presidential candidate, who was certainly expected to have served conscientiously on the Council for some years or at the very least on various committees. And with improved communications residence in London ceased to be a requisite for officers, so notably that from time to time there were to be accusations of a Cambridge clique, just as early in the century there had been accusations of dominance by the Royal Institution or the British Museum. By the last quarter of the century the officers indeed were virtually all university-based, a far cry indeed from the situation in the first quarter of the century.

Although the power of the President was diminished, the fact that the office now appeared to place its incumbent at the head of English science

Fig. 10. William Parsons, 3rd Earl of Rosse (1800–67), F.R.S. 1831,
P.R.S. 1848–54.
By J. Catterson Smith

made it highly desirable and prestigious, demanding both scientific eminence and worldly ability. Hence there was always a certain degree of canvassing for successors, although never on the scale of the past. Further, in the absence of regulations the tenure of each President was at his own discretion; although after 1871 no President held the office for more than five years, this was never statutory. Successive Presidents naturally held differing views about the rôle which the office demanded, but all saw themselves as at the head of British science.

The Presidents: the astronomical lords 1848–58

Rosse's election had gone through smoothly, without any overt dissension, but his choice involved a decided break with the past. There had not been an astronomer President for eighty years, and Rosse, not yet fifty when elected, was still, as noted above, an active, practising scientist – a fact indeed which made it awkward for him to attend Royal Society functions during the autumn and early winter, when he needed to be at home in Parsonstown (Birr) to pursue the researches with his great telescope which led to his discovery of spiral nebulae. He was, to be sure, an active member of Parliament like his three immediate predecessors, but Parliament did not then sit in the autumn, so that, as he said, the London season for him was naturally the spring. He had not been particularly active in the Royal Society's affairs, nor in those of the British Association for the Advancement of Science, although he had willingly served as its president in 1843 when it met in Cork. Now he proved himself a conscientious administrator, who served as President of the Royal Society for six years (still regarded as a relatively short term of office) although he was not entirely happy with his rôle.

His dissatisfaction had two sources. A minor one, already mentioned, was the Royal Society's calendar. Although he faithfully attended the Anniversary Meeting each November during his term of office, he would, like Northampton before him, have preferred to have Presidential formalities all concentrated after Easter, at the height of the London season, and indeed established the tradition of holding the soirées then, although he never succeeded in shifting the Presidential address to June, as he would have liked to do. His other, more serious source of dissatisfaction was his lack of control over the Council, which must in part have stemmed from his absence from Council meetings during the autumn and early winter, although it was implicit in the reforms of 1847. The reformers had made no secret of the fact that they believed that the Royal Society should be almost exclusively composed of practising scientists, and the Council entirely so, while Rosse, more worldly wise and also more concerned with relations with Government (see Chapter 6), saw the matter otherwise. His first defeat came in June of 1849, only a few months after his election, when George Rennie, who had served as Treasurer since 1845, expressed his intention of resigning, as he was finally to do a year later under the stress of financial embarrassment;[1] Rosse then announced his proposal to appoint Sir Robert Harry Inglis (F.R.S. 1813, an extremely conservative Tory politician, elderly but influential, whose intellectual interests were those of an antiquary) to succeed him. When Rennie, perhaps as a consequence, withdrew his resignation a fortnight later, Horner (as usual taking a 'radical' line, that is one favouring scientific rather than worldly Fellows) threatened to propose a motion that 'it is expedient that the gentleman recommended by the Council for such office [as that of Treasurer] should have actively engaged in the pursuit of science', on the grounds that the Treasurer was senior Vice-President and on all scientific and publication committees. The threat was enough: Rosse could do nothing but regret that Inglis, whose influence he valued, was dropped from the Council in November 1849 while Rennie was retained on the Council after being replaced as Treasurer by Sabine in 1850.

Rosse's frustration emerges clearly in his letters of 1854 which notified the Council of his views on the future composition of the Council and his dissatisfaction with his past relations with it. There he stressed what he saw as the need on the Council for men with influence in society and Government, who knew how to approach civil servants and cabinet ministers and how best to influence them. While radical reformers had for decades claimed that the Government ignored science because the Royal Society was not exclusively scientific, Rosse thought that the Council should not only be fully representative of the whole range of science but should also contain a number of men 'conversant with worldly affairs' so that the Society might carry 'political weight' – and if that meant a larger Council he was prepared to propose a change in the charter to that end.[2] Indeed, as he told the Council when resigning in 1854, he differed from them on a matter of '*principle*' for he thought that while the Council should be '*largely*' composed of men devoted to science, these should always be 'chosen mainly in consequence of their position in society', men who 'might be likely to influence Her Majesty's Government' hoping thereby to 'relieve my successor from the disagreeable position of being the nominal head of English Science but really in my case little more than a puppet'.[3] Sabine in replying was emollient, but firm.[4] He first pointed out that the

Council had so far acceded to Rosse's point of view as to substitute the name of the Earl of Harrowby (F.R.S. 1853) for that of Lyon Playfair (F.R.S. 1848), for the members saw the advantages of appointing a man who had been associated with Rosse and Wrottesley 'in the proceedings of the Parliamentary Commission' [of the British Association], was influential in Government and had been president of the British Association at its Liverpool meeting. But he insisted that others suggested by Rosse were not suitable: the Duke of Northumberland (whom Rosse had favoured as President) had been elected to the Council in 1847 but had attended only once; the Duke of Argyll (F.R.S. 1851) had indeed shown an interest in old-style (non-evolutionary) geology and ornithology and had been president of the British Association, but as a cabinet minister (he was Lord Privy Seal, about to become Postmaster General) he could not be expected to attend Council meetings at all frequently; Lord Ashburton (W. B. Baring, also a Whig, F.R.S. 1854) was too recent a Fellow; and as for Sir Robert Inglis (whom Rosse still favoured) the Council, as Sabine assured Rosse, recognised his qualities but rejected him as a member on account of the fact that 'in late years' there had been an increase in those 'duties' of the Council requiring '*special* scientific knowledge in its members'. At the same time Sabine reported that the Council had learned with '*very great regret*' that Rosse himself did not wish to continue on the Council (as was normal practice) and rejoiced that it was already too late to remove his name from the slate.

Rosse saw this difference in point of view as a matter of 'principle'. The real difference was in what was viewed as the Council's main function, particularly in regard to the Government Grant (begun in 1849; see Chapters 6 and 7). They agreed about its importance, but while Rosse thought that influential men were required on the Council to keep the Grant an annual affair, that is to retain the interest of the government ministers, Sabine thought that the Council needed scientific men for the effectual administration of the Grant, that is, the choice of recipients. It was possibly in part a difference arising out of a difference in social background; certainly Rosse, though the younger man, had a more old-fashioned point of view, seeing Government in terms of men and influence, while Sabine ignored the need to keep in personal touch with the men who were the Government, concerning himself entirely with administration and with the civil service departments. Slowly indeed a line of demarcation between Council and President was beginning to appear, so that although successive nineteenth-century Presidents were all to be men of high scientific attainments, most were men whose careers had given them a prestige not confined to academic scientific circles. The President was now a constitutional monarch.

Rosse and the Council readily agreed on his successor, Lord Wrottesley. Indeed, his interests were so similar to Rosse's own that it almost seems as though the Council, far from wanting to restrict Presidential terms (although this had been so often suggested) thought six years of office too short and so chose an almost indistinguishable successor to serve for the next four years to make an even ten years – Northampton's period in office earlier. Sir John, second Baron Wrottesley, only two years older than Rosse, like him an Oxford graduate, was also an astronomer, who had been active in the foundation of the Astronomical Society of which he had been successively secretary (1831–33) and president (1841–43). He was F.R.S. 1841, shortly after succeeding to the title. Two years before that he had tried to secure astronomical posts at Oxford in succession to S. P. Rigaud (G. H. S. Johnson, F.R.S. 1838, was in fact successful), but he was wealthy enough to maintain two observatories, one at Blackheath and one at Wrottesley (Staffordshire) and was a serious astronomer of professional standards.[5] He had served faithfully on the Royal Society's Council on a number of occasions, was active in the British Association (especially on its Parliamentary Commission of which he was founder and chairman; he was to be president at the Oxford meeting of 1860), and active in Parliament, serving on numerous Royal Commissions. He was altogether a choice after Rosse's own heart as a man of affairs, but possessed of scientific attainments to suit the Council. He was said not to be popular,[6] but his letters reveal a modest man: thus in 1847, writing to press Herschel to serve on a Royal Commission investigating the collapse of a cast-iron bridge over the River Dee, he wrote that he was to be Chairman 'for no other reason I suppose than because my name must stand first in the list' and Rosse reported that he felt himself unworthy of the office.[7]

As P.R.S. he was conscientious and reasonably effective, more faithful in attendance than Rosse, and carried on successfully the important developments begun under the latter's Presidency. Thus when in 1854 – five years after initiating the important Government Grant – the Treasury suddenly announced that it did not intend to continue the Grant, which the Society had taken to be an annual one, Wrottesley handled the matter in a quietly able manner.[8] Although the innovation of the Government Grant belongs to Rosse's Presidency, its firm establishment should be credited to Wrottesley (see further Chapter 6).

Similarly, it was Wrottesley who completed the removal of the Society's premises from Somerset House in the Strand to Burlington House in

Fig. 11. The Meeting Room in Somerset House. From a contemporary engraving

Fig. 12. The First Meeting Room in Burlington House (1863). From a print in the *Illustrated London News*

Piccadilly. This had been rumoured as early as 1847,[9] when the library had begun to outgrow its rooms; further Burlington House was closer to the shifting centre of social and intellectual life (it already housed London University) and to Parliament. The new rooms were occupied for the first time early in May 1857, a week after the last meeting in Somerset House (30 April).

That the Society had achieved some stability, which Wrottesley's Presidency continued, is demonstrated by the unsuccessful attempt of the irascible James South to stir up old quarrels and to foment discontent over policy.[10] South had two grievances: one, the major, was rather to do with his personal affairs than with public ones, though it was occasioned by the obituary notice of Richard Sheepshanks (F.R.S. 1830) printed in the *Proceedings* for 30 November 1855, for South had an old quarrel with him; the other was an attack on the Royal Society for not accepting Rosse's proposal of an enlarged Council. In fact South's view was very different from Rosse's; he insisted that the Royal Society's Councils were insufficiently scientific, for, as he described it, they

> generally contain a certain number of titled or eminent persons,
> whose avocations preclude them from rendering proper attention to
> the scientific and ordinary business of the Society. The result is,
> that *little men*, under the shadow of this aristocracy, have a
> commanding influence over the proceedings – elect their own
> miserable *toadies* to office, and do everything in their power to
> crush men of learning who will not be subservient to their selfish
> views and absurd pretensions; in fact, the working and unassuming
> men of science are insulted and oppressed by the *little men* under
> the name of 'The President and Council of the Royal Society'.

Rosse must have been surprised by South's advocacy, to say the least. In fact South showed himself thoroughly out of touch with the Society's present state and even the press showed no interest in his pamphlet.

The elder statesmen 1858–73

Wrottesley's departure from the Presidency was as quiet as his arrival had been. No correspondence survives; the minutes of the Council (6 May 1858) merely state 'Lord Wrottesley intimated his desire not to be put in nomination for the Office of President at the ensuing Anniversary'; and unlike many of his predecessors he said nothing in his Anniversary address about either his past or his successor's future (White reported that 'awkwardly enough [he] did not induct [his successor], but left him to install

himself in the Chair'.) Debating the matter of a new President a fortnight
after Wrottesley's notification the Council agreed to ask Faraday, and a
deputation consisting of Wrottesley, Grove and Gassiot went to the Royal
Institution to ask him if he would be willing to serve.[11] Although only two
years before Faraday could say 'I have not been able for several years to
occupy myself with the Royal Society or its management' (and, indeed,
although a frequent and faithful member of the Council in the 1830s he had
not served on it since then, although he had often acted as a referee of
papers),[12] he was clearly tempted by the opportunity to improve the Society,
and the honour, but in the end decided against accepting the offer. As John
Tyndall recollected ten years later, Faraday told him, after considerable
reflection and discussion,

> I must remain plain Michael Faraday to the last; and let me now
> tell you, that if I accepted the honour which the Royal Society
> desires to confer upon me, I would not answer for the integrity of
> my intellect for a single year,

in which he was probably right, for his health, both mental and physical,
was now, in his sixty-seventh year, precarious.

His supporters were disappointed but accepted his decision, turning
instead to Sir Benjamin Collins Brodie, the first surgeon to become
President of the Society, an eminent man who was presumably chosen to
give the biological side representation again after a long gap. It is perhaps
a measure of confidence in the stability of reform that Brodie could be
chosen, for he was elderly (b. 1783), had been elected F.R.S. in 1810, well
before the end of Banks's time, and his chief physiological work (for which
he had received the Copley Medal in 1811 and which had served as a basis
for the Croonian Lecture in 1813) was long in the past, his physiological
researches having ended in the early 1820s with his lectures to the Royal
College of Surgeons.[13] He had since been physician to three successive
sovereigns, been created baronet in 1834, been president of the Royal
College of Surgeons, and was D.C.L. Oxford, 1855, thus having come a
long way from the days when the editor of *The Lancet* could refer to 'little
Brodie'.[14] As President he was faithful at Council and ordinary meetings
for two years, until his sight began to fail, leading to his resignation in 1861,
a year before his death.[15] His contribution to the Society was thus
necessarily slight, but his advocacy of open discussion, in both Council and
ordinary meetings, was surely beneficial.

Although Brodie's three-year Presidency was peaceful, his resignation
caused problems. Two candidates were initially considered: Thomas

Fig. 13. John, Baron Wrottesley (1798–1867), F.R.S. 1841, P.R.S.
1854–58.
Photograph by Maull & Co.

Fig. 14. Sir Benjamin Collins Brodie, Bart (1783–1862), F.R.S. 1810,
P.R.S. 1858–61.
Photograph by Maull & Polyblank

Graham and Lord Brougham.[16] Graham (F.R.S. 1836) was the leading contender, and a natural one, a distinguished chemist who had received a Royal Medal in 1850 and was to receive the Copley Medal in 1862, professor of chemistry at University College London until 1855 when he resigned to succeed Herschel as Master of the Mint, first president of both the Chemical Society (1840) and the Cavendish Society (1846). He was clearly tempted, and indeed apparently almost accepted first the Presidency and then, after withdrawing from that honour, the Treasurership, in the end rejecting both on the grounds of ill health – although not yet fifty-six in the autumn of 1861 he was not well, and when he died eight years later it was thought to be from overwork. That he regretted his final rejection is suggested by the rumour which reached the Assistant Secretary, Walter White, on 5 November 1861 – 'I hear that Mr. Graham wished he had accepted the Presidency' – although the soundness of this rumour is rather contradicted by White's addition 'and that Mr. Faraday says if it were offered to him now he would accept it', for, at seventy, Faraday felt his powers waning and was to retire completely from the Royal Institution the next year.

Another candidate, favoured strongly by Roget and presumably by some others, was Lord Brougham (F.R.S. 1830), former Lord Chancellor, a founder of the *Edinburgh Review* and of London University, a great and, on the whole, successful reformer, who had possessed some scientific competence in his youth and had revived this interest in old age (he was, in 1861, eighty-three, and was to live to be ninety). Roget's reasons for supporting Brougham were described by Charles Wheatstone (F.R.S. 1836, professor of experimental physics at King's College, London, where his work on electricity, acoustics and optics was carried out) as follows:[17]

> Dr. Roget told him [Wheatstone] that he (Dr. R.) was one of the prime movers in the question of having Lord Brougham as P.R.S.; that there are many reasons why his lordship should be preferred to General Sabine, ignoring the fact that by his sneer at Young's Undulatory theory in the 'Edinburgh Review' he disgusted Young, who quitted the subject, and Fresnel took it up, and the honour of the discovery went to France; that there is nothing really new in his Lordship's optical papers, that the results are all calculable, and that the Académie reads his papers but does not publish them.

(He was never elected a member or correspondent of the Institut.) Wheatstone's view of Brougham's scientific attainment was no doubt shared by others. On the other hand support for him was very real, as Brodie's

Anniversary address suggests, and as a letter from the Reverend James
Booth (F.R.S. 1846) testifies. He declared[18]

> I understand that a requisition to Lord Brougham is in course of
> signature among the fellows of the Royal Society to allow himself
> to put in nomination for the office of President when ever the next
> vacancy should occur.

and indicated that he would like to sign such a requisition. (He was clearly
not in close touch with events, which had already passed him by.) The
problem had been stated most clearly a year earlier on Brodie's resignation
by W. H. Miller (F.R.S. 1845, professor of mineralogy at Cambridge) who
wrote to Sharpey[19]

> I am in dismay at the prospect of a search after a new President,
> and I read over the list of Fellows again and again without deriving
> much consolation from the task. I believe General Sabine would
> make the best possible President but how far this opinion is shared
> by yourself and other influential Fellows of the Society I know not.
> Still less do I know whether this view prevails with the Society at
> large to such an extent as to keep out any rival Candidate. The
> advantage of having an excellent President instead of a moderately
> bad one would not counterbalance the evil arising from a contest.

Contrary to his doubts, Miller's opinion *was* shared by the Council, which
on 20 June 1861 decided to offer the nomination to Sabine who, tactfully,
had absented himself, contrary to his usual custom, showing that he hoped
for the nomination.

General Edward Sabine at the age of seventy-three was still extremely
vigorous (he was to live another twenty-one years), sufficiently distinguished
in his profession (at this time a Major-General, promoted Lieutenant-
General 1865, K.C.B. 1869 – a slightly Gilbertian soldier who had not been
on active service since his very early years), and with a long association with
the Royal Society. He had been elected a Fellow in 1818 when a young
artillery officer, as an explorer and embryo geophysicist, had served as
Secretary 1827–30, Foreign Secretary 1845–50 and Treasurer 1850–61. He
had thus lived through all the period of great reforms, voting on the whole
in their favour, and as he had always been assiduous in his duties and in
attendance at Council meetings (in addition to his terms of office he had
been a member of the Council 1839–40) he was fully conversant with the
Society's affairs. Sabine stood for harmony and cooperation, as had been
exemplified in his activities within the British Association, of which he had
been general secretary from 1839 to 1859, in his lifelong devotion to the
collection of magnetic data and even in his abhorrence of 'speculation', as

he called the construction of any hypothesis or theory. His one set-back had been the accusation by Babbage in 1831 that he had falsified his data; his later apologists denied the charge, but admitted (as seems to be indeed the case) that he was incapable of properly sophisticated numerical analysis. His forte lay rather in the promotion and organisation of the collection of data, especially magnetic data, by others. His early reputation as an explorer and his later rôle as a leader in geomagnetic research combined to raise his scientific reputation to a respectable level, and he was an effective scientific administrator, well in tune with his times, while Babbage, who had chosen to play no active part in affairs of which he disapproved, was left petulantly campaigning for a cause now obsolete. The scientific generation which had entered the Royal Society since the great scandals of the early 1830s (like Gassiot and Grove, both F.R.S. 1840, or Spottiswoode and T. H. Huxley, 1850 and 1851) and who took the phrase 'The President and Council' literally, with the Council dominant, found Sabine an acceptable figure to preside (not, as in former days to rule) over the reformed Council and Society of the 1860s even though they did not entirely trust him.[20] Indeed he was to make an amiable, pleasant and sufficiently popular P.R.S., who rather touchingly in his final Presidential address thanked the Fellows for having supported the soirées he had provided (at his own expense, as a matter of course).[21]

The changed relation of President and Council meant far more work for the Council, as was widely recognised,[22] but for a conscientious and zealous President like Sabine, who attended virtually all Council meetings and most of the meetings of the committees of which he was *ex officio* a member, it was a time-consuming responsibility. Sabine clearly devoted a great deal of his time to the post, and unlike younger men found no difficulty in so doing. For a number of years all went smoothly: Sabine, familiarised with the Society's needs by his earlier years in office, worked harmoniously with the officers and Council, endeavouring to avoid any possibility of friction. He appreciated the work of the officers, including the salaried assistants: in 1868 the Assistant Secretary, White, recorded that 'In talk this afternoon our President said one principal reason why the Royal Society stood better than ever in public opinion, was the diligent way in which the business of the office was carried on'.[23] But he saw that they needed guidance. Thus in 1863 he could caution Sharpey (senior Secretary), 'Last year we had *six* Naturalists on the Council; and not a single Astronomer...we must not continue so great a disproportion, or there will be complaints not without reason at the Anniversary';[24] while the next year, when expressing pleasure at the nomination of Gassiot to the Council (which he had not urged because Gassiot was his close friend), he wisely noted, 'I think moreover that three

Cambridge professors ought not to be [on the Council] without *one at least* Oxford Professor.'[25]

But in time friction did develop, and on two counts. The first was that Sabine, inevitably, retained some of the scientific judgements of his youth and was not in sympathy with the views of younger men: thus while accepting the Council's decision in 1864 to award the Copley Medal to Darwin he managed in his Presidential address to exclude Darwin's evolutionary work from the grounds of the award, to Huxley's great and natural indignation, an indignation so general that Sabine modified the offending phrase in the published version. The second basis for friction was administrative: as time went on Sabine tried to resist the Council's dominance. So in 1868, when he had been President for seven years, he complained to White of the Council's preparation of the house-list of officers and Council, which he wanted prepared by the officers for the Council's possible emendation, declaring 'That he would not be President to fight a faction, and that he intended to resign soon'.[26] A year later he appeared to be reconciled; for when A. B. Granville, one of the reformers of the early 1830s, wrote wistfully that in over fifty years as a Fellow he had never been on the Council and would like to be so, Sabine replied[27]

> I believe that at the time when you and I were elected Fellows, the selection [of the Council] rested entirely with the President...; but the practice is very different now and has been so for several years...At a meeting of the Council next before the day on which a nomination is to be made, each member of the Council is requested to prepare and place in the hands of the Senior Secretary a list of Fellows for the next Council. These lists form the foundation upon which the Officers, acting as a Committee of Selection, propose what for distinction's sake may be regarded as a House-List, which is submitted for consideration at the next meeting of the Council which is specially summonsed for this duty. At this meeting many changes are frequently made either by proposition and general assent, or by distinct motions made by individual members and decided upon by a majority of voices. It is of course a guiding Principle that all the Sciences which are regarded as being within the province of the Royal Society, should be adequately represented; and so far as I am able to judge this principle is pretty generally kept in view. But it must occasionally happen that subjects are known to be likely to require the consideration of the Council in the ensuing year, which may make the presence of a certain Fellow, or Fellows, very desirable; in such

cases it is at the discretion of the Officers to place such name or names upon the House-List...*at the present time*, whatever it may have been formerly, the President can exercise no other influence on the nomination of the Council than such is open to any other of the Members of the Council.

However, although Sabine had partially accepted the Council's dominance, he refused to accept its judgement on the optimum length of a Presidential term of office, a matter long controversial, although quiescent during some years because successive Presidents (Rosse, Wrottesley and Brodie) had all resigned spontaneously. Sabine toyed with the idea of retirement in 1868 out of pique, but showed no real inclination to take the decision, even in his eightieth year. Not until October 1870 did he intimate to the Council that he would not serve more than another year, and then under provocation.

The affair emerged into the open and is described at length by Sabine's intimate friend Gassiot, who reverted to earlier custom in producing a privately printed pamphlet 'for private circulation'.[28] This is a defence of Sabine against the accusation that he had only resigned because the Council refused to appoint Gassiot (himself over seventy) as Treasurer. As so often happens, the argument is confusing, now that the events are long past; what does emerge, confirmed by letters, is that, as Gassiot put it,'a desire for a change in the Presidentship had manifested itself for some time previously', adding 'Dr. Sharpey, the Senior Secretary, and the late Dr. Miller, the Treasurer, had repeatedly urged the writer to speak to the President on the subject of his retiring from office, which assuredly they would not have done had not such an unusual course suggested itself to them from outward pressure.' Gassiot claimed to have refused, until he discovered the depth of the opposition to Sabine's continuance, but then Sabine was obdurate, believing that he still had work to do. This was perhaps in 1868; by 1870 Sabine was more seriously thinking of resigning, as he wrote to Gassiot, but not until the Council meeting of 27 October did he make a public statement, an action effectively forced on him by A. W. Williamson (F.R.S. 1855, professor of chemistry at University College London) who brought to the Council's attention a resolution of Council on 13 April 1848 that it was the Council's opinion that it was 'inexpedient' that the same man should be nominated as President by the Council 'for more than four successive years', a resolution much applauded by the reformers but never yet rendered effective.[29] Others besides Gassiot thought Williamson's proposal inappropriate to say the least; Sabine told White that 'The Marquis of Salisbury expressed himself strongly on the rude scene in the

Council of 3 November, and Mr. Mallet [F.R.S. 1854, civil engineer] wrote that the phrase "infusing new blood" meant infusing those who used the phrase'.[30]

Sabine felt himself 'not well treated' to have his suggestions for Council members (Owen and Wheatstone) rejected, and he hoped for a noble successor: in 1868, according to White, he had approached the Duke of Argyll; in 1870, according to the same authority, he suggested Lord Salisbury (F.R.S. 1869), then still a young and active extreme Conservative, chancellor of Oxford, with some interest in science, to serve for a couple of years, after which Rosse might be again offered the office. But he vacillated; he also 'spoke of Mr. Grove as one who might make a pretty good President: but that he hears he has become too fond of sneer and snarl'.[31] There is no indication that Sabine had anything to do with the final choice of his successor, and indeed he was not in the chair at the meeting of 19 January 1871 when Joseph Hooker (F.R.S. 1847, now director of Kew Gardens) chose to raise the matter by declaring that he intended to ask the Council to consider the 1848 resolution. (This he did, but a modified proposal effectively to limit the Presidency to five years was not, after discussion, pressed home.) Perhaps this had something to do with the choice of G. B. Airy (F.R.S. 1836), the first Astronomer Royal, to serve as President.

Airy, who was seventy when he took office, had been successively Lucasian and Plumian Professor at Cambridge as a young man before becoming Astronomer Royal in 1835, F.R.S. 1836; as Astronomer Royal he enormously added to the prestige of Greenwich Observatory and had a long association with the Royal Society both in his official capacity and as a member of the Council on a number of occasions. He was highly regarded as an astronomer, especially for his work on planetary perturbations,[32] and for his organisational ability well displayed in his autocratic regime at Greenwich. Besides being the first Astronomer Royal to be chosen P.R.S., Airy was probably the first President ever not to be wealthy when he took office: he had previously refused all honours except a pension of £300 a year from Peel in 1835 on the grounds that he would not be able to afford to keep up the position requisite should he become a knight (an honour offered to him in 1835, 1847 and 1863; he finally accepted a K.C.B. in 1872). This view of his was well known and explains why in March 1871 the Council decided that in future the costs of Royal Society soirées should no longer be defrayed by the President but by the Society, an action which led to the regular annual appointment of a Soirée Committee; only after that decision did the Council approach Airy, who accepted, but with some reservations

Fig. 15. General Sir Edward Sabine K.C.B. (1788–1883), F.R.S. 1818, Secretary 1827–30, Foreign Secretary 1845–50, Treasurer 1850–61, P.R.S. 1861–71.
Photograph by Wilson & Beadell

Fig. 16. Sir George Biddell Airy K.C.B. (1801–92), F.R.S. 1836,
P.R.S. 1871–73.
Engraving by I. W. Slater from a drawing by Wageman

about the expenses in which he would be involved (which did not create a favourable impression. He was reported to be asking for travelling expenses, and Sharpey was said to have commented drily 'with a cab to and from Charing Cross, and his house and Greenwich rail, he might solve the difficulty'.)[33] Airy's own private comment was[34]

> The election to the Presidency of the Royal Society is flattering,
> and has brought to me the friendly remembrances of many
> persons; but in its material and laborious connections, I could well
> have dispensed with it, and should have done so but for the
> respectful way in which it was pressed on me.

Airy was certainly regarded with respect, and he presided with dignity, but he was not a man who could be an effective or memorable President for he was too deeply involved in his own scientific concerns. He himself recognised the fact, and soon announced his intention of resigning after only a two-year term. In his final Presidential address[35] he was to comment

> I respect the sentiment which has prompted the Society to seek for
> its President a man of supposed scientific character, and, perhaps in
> preference, a man in official scientific position, and I join in the
> unanimous feeling of the Council that, this principle being
> admitted, its application could never have been better made than in
> the selection of the Fellow whom they recommend to you as
> successor to myself. But I still think that, practically viewed, the
> principle is not the best that can be adopted – and that
> consideration on the leisure which the President can devote to the
> concerns of the Society, on the proximity which enables him at any
> moment to enter into its business, and on the personal vigour
> which he may be expected to bring into all his transactions with it,
> ought to hold a very important place.

(He had already told the Fellows that he was 'grieved' not to have devoted more time to the Society's affairs.)

Although the Society was to continue to choose as President Fellows of eminent scientific reputation, these were never again to be so elderly as Airy, none being seventy or over on election, and few so elderly even on the completion of their terms of office, which from this time forth were never to exceed five years. This, and the change in arrangement of the soirées, from being the President's sole responsibility, financial and social, to rather being the joint responsibility of the President and Council, as with most of the Society's affairs, were the most significant effects of Airy's Presidency.

A less fortunate effect of the shift of responsibility was that any gossip appearing in the press had the air of being issued officially, and some such gossip was inevitable, given the public interest; but on the whole it was contained by quiet secretarial action.[36]

The new men 1873–85

The nomination of Joseph Dalton Hooker in succession to Airy was something of an innovation. He was the first botanist since Banks. Still relatively young (born 1817, F.R.S. 1847) he belonged to the new class of second-generation, professional scientists: he had travelled extensively, having been on Ross's antarctic expedition from 1839 to 1843 and then, a real novelty, in the Himalayan regions from 1848 to 1849; he had succeeded his father as director of Kew, its gardens and research establishment, in 1865; and although elected F.R.S. before the 1847 reforms, had always favoured the 'radical' point of view. He had been an active member of the Philosophical Club, founded in 1847 to discuss and promote reform, and of the *x* Club. The latter,[37] founded by Huxley in 1864, consisted of nine friends who dined weekly (usually at the Athenaeum) before the Society's meetings, eight of them being F.R.S. (The exception was Herbert Spencer, who was never proposed.) Besides Hooker, both Huxley and Spottiswoode were to be P.R.S.: three others (George Busk, surgeon and naturalist, F.R.S. 1850; T. A. Hirst, mathematician, F.R.S. 1861; Edward Frankland, chemist, F.R.S. 1853) were active in the Royal Society and in the British Association, while John Tyndall (of the Royal Institution, F.R.S. 1852) and Sir John Lubbock (son of the former Treasurer, best known as an anthropologist and man of affairs) were both presidents of the British Association. The *x* Club was a device to permit busy men to meet their friends, but they naturally discussed the affairs of the Society to which they were devoted, and often agreed informally on candidates.

Hooker's positive qualifications were his record as an able administrator, his undoubted scientific attainments, and his personal character. What Sabine had written of him in 1856 – 'He...has seen and thought much: is thoroughly unbiased by personal or party motives; and can be depended upon for regular attendance'[38] – suggested dependability rather than brilliance. Hooker himself in 1872, passing on to Darwin what he regarded as a great compliment, that the two friends had recently been described as the two most modest scientists of their time,[39] had declared

> I quite agree as to the awful honor of P.R.S., and its inestimable
> value to me in my position, and under existing circumstances – but

Fig. 17. Sir Joseph Dalton Hooker, O.M., K.C.S.I. (1817–1911),
F.R.S. 1847, P.R.S. 1873–78.
Photograph by Maull & Co.

Fig. 18. Thomas Henry Huxley (1825–95), F.R.S. 1851, Secretary
1872–81, P.R.S. 1883–85.
Photograph, 1877

my dear fellow, I don't want to be crowned head of science. I
dread it – 'Uneasy is the head,' &c...

He had indeed refused a knighthood (K.C.S.I.) earlier, at the time of his
C.B., on the grounds that he cared nothing for a title unless it were clearly
seen to be an award for *scientific* achievement. (He finally accepted a
knighthood in 1877.) Not everyone approved his modesty, for many
Fellows liked to have the world see that their President was a distinguished
man; some outspokenly preferred a nobleman and the Duke of Devonshire
was again suggested.[40] Besides, although Hooker's reputation as a radical
(in the Society's affairs, in science, and in consequence of his close
friendship with Huxley and Darwin) was a recommendation to some, it was
a disadvantage in the eyes of others; the Council gave him a three-fourths
majority, not the formal unanimity which Sabine's nomination had received
eleven years earlier.

As expected, Hooker proved a conscientious and able President. Well
aware of the doubts about himself and about the undesirability of long terms
of office he made it a condition of acceptance that he should hold the office
only from year to year,[41] in fact retiring after a five-year term (henceforward
customary) in 1878. He worked hard; on Council days this meant continuous
meetings from 1 p.m. to 6, then dinner, then the regular meeting at
8.30 p.m., while on other days there were committee meetings, both at the
Society and at the British Museum of which the President was still *ex officio*
a trustee. As he told Darwin 'I have 15 Committees of the R.S. to attend
to', but, especially in 1875, after his wife's sudden death, he found it all
a pleasant change from his usual business at Kew. As he put it,[42]

> I cannot tell you what a relief they are to me – matters are so ably
> and quietly conducted by Stokes, Huxley and Spottiswoode that to
> me they are of the same sort of relaxation that Metaphysics are to
> Huxley. I have no sense of weariness after them. Of course I must
> expect some rows and difficulties in the Society, and they will come
> when least expected, you will say – but meanwhile let me enjoy my
> illusions.

In fact Hooker seems to have been spared major storms. There continued
to be those who were unhappy about the rejection of their papers,
occasionally to the point of resigning, but on the whole the Secretaries dealt
efficiently with such matters, as they did with complaints in regard to the
conduct of Fellows or of potential Fellows, and only occasionally did the
President have to intervene.[43]

In his capacity as President Hooker tried to keep the Fellows informed

All scientists now

of the state of their Society and of its current concerns. As he told Darwin
in the autumn of 1874, when he was engaged in preparing for his first
Anniversary Meeting,[44]

> I am busy with my address for R.S. which I am advised to make a
> purely business one, and to confine it to the operations of the
> Society, its Committees, Funds, labours under Government and
> private affairs, about which it appears that the Fellows in general
> are absolutely ignorant. They know nothing of the Donation Fund,
> Government Grant, Sc. Relief Fund, and the dozen or so
> Committees, many of them Standing Committees, that involve an
> amount of work on the part of the officers that not only justifies
> paying the Secretaries but makes it expedient for the Society to do
> so, and necessary to support themselves,

and indeed his address is a most useful summary of the then mode of
working in the Society. By this time the secrecy surrounding the affairs of
the Society was much diminished, lists of candidates nominated by the
Council and the November slate of officers being regularly sent to the
'Public Papers' – usually *The Times*, *The Athenaeum* and *Nature* (founded
at the end of 1869). Only in times of crisis were such journals, or *The Lancet*,
likely to resort to comment, criticism, attack or gossip.

An unexpected result of Hooker's first Presidential address was that a
Fellow not on the Council (Dr C. G. H. Williams, F.R.S. 1862, a chemist)
raised the question of the wisdom of limiting the number of Fellows; an
Election Statutes Committee was appointed, but it concluded that there was
no reason to change, as Hooker duly reported to the Anniversary Meeting
in 1875;[45] the eventual separation of the *Philosophical Transactions* into two
series, A (mathematical and physical sciences) and B (biological), mentioned
by Hooker in 1874, was slowly proceeded with, to be finally effected in 1887.
In 1878 Hooker was able to inform the Society of the successful inception
of a scheme of his own devising, a roundabout method of reducing the fees
paid by Fellows which had been long felt to be so excessive as to deter some
able men from becoming candidates. As he himself explained the matter,[46]

> I have long had at heart a scheme for reducing the monstrous
> heavy fees (in future) of F.R.S. by establishing a 'publication
> fund', which by relieving the income of part of the expenditure on
> publication, would eventually set free the desired amount for
> reduction of fees to the standard of other Societies,

for as he noted in his Presidential address that year, the fees were far higher
than those of other societies. The Treasurer (Spottiswoode) calculated that

a fund of £10,000 would suffice; this was quickly raised in large part by Hooker's canvassing personally, and had the desired effect, for the admission fee of £10 was abolished, and the annual subscription reduced from £4 to £3.[47] All in all, Hooker had amply demonstrated that an eminent scientist and scientific administrator, still active in his profession, could be an effective and able President although possessed of no social or political influence, especially when, as in his case, assisted by a generous, warm personality.

Hooker offered his resignation to the Council at the meeting of 21 February 1878, on the grounds that he was otherwise too much occupied to serve effectively; another motive may well have been the desire to institute the practice of five-year Presidential terms. By mid-May the Council had quietly decided to nominate Spottiswoode. William Spottiswoode (b. 1825, F.R.S. 1853) belonged firmly to the post-reform generation. He was an Oxford graduate, elected F.R.S. for his work in pure mathematics; he also worked in optics and electricity and became a successful lecturer and writer on these subjects, while actively engaged in business – he had succeeded his father as Queen's printer and partner in the publishing firm of Eyre and Spottiswoode in 1846. He was thus wealthy; he was also a close friend of Hooker and his circle, a member of the x Club. He had twice served on the Council before he was chosen Treasurer of the Society in 1870, with no real opposition (although Sabine had preferred Gassiot, it was not because he disliked Spottiswoode). He was president of the Mathematical Society 1870–72 and of the British Association at its meeting at Dublin in the summer of 1878. He was thus a natural choice, although in retrospect scientifically less distinguished than either his predecessor or his successor, and provided a President on the A-side after Hooker's representation of the B-side, and, moreover, one conversant, like Sabine, with the financial responsibilities of the Society.

Like Hooker, he began his Presidency by surveying the state of the Society for the benefit of the Fellows in his first Address (1 December 1879): he noted changes to statutes to reduce annual payments and abolish the entrance fee; a new procedure for papers, which were now surveyed in sufficient time to allow a list to be, as he put it, 'advertised for reading' – in the 'weekly Journals' – which also facilitated quick publication in the *Proceedings*; and a better abstracting service by the Secretaries which had led to a partial circulation in advance. He then suggested that the Fellows consider a change in meeting hour to 5 p.m. to permit a return home by 7, with the post-meeting tea to be replaced by afternoon tea; a year later he could announce that the time had in fact been set at 4.30 (as today) with the Council meeting fixed for 2 p.m. He proposed the installation of electric light, which would assist the optical illustration of papers as well as being generally useful, and announced that he was offering the Society an 8 h.p. gas engine, while Siemens was offering 'a pair of dynamo-machines'.

On one matter of government Spottiswoode appears to have differed from the views of his immediate predecessors: he does not seem to have intended to serve for any very short term; certainly he did not intend to resign in November 1883. That his term came to an end prematurely was the result of the still unsatisfactory state of England's water supplies and of medical knowledge: like Prince Albert some twenty years before him he died of typhoid fever, on 27 June 1883. A hastily summoned Council meeting (5 July) unanimously elected Huxley as President *pro tem*, a decision to be confirmed in November when the slate for the coming year was determined.

The choice of Huxley seems in retrospect a very natural one. He had been elected F.R.S. in 1851, at the age of twenty-six, for the zoological research undertaken during his term as a naval surgeon on the *Rattlesnake* during its survey of Australian waters; as he wrote triumphantly, he had been lucky in this success, or as he put it,[48]

> without canvassing a soul or making use of any influence, I have been elected into the Royal Society at a time when that election is more difficult than it has ever been in the history of the Society.

Sabine had encouraged him to stay in England as a naturalist rather than return to duty in the navy; he became very friendly with Darwin and with Spottiswoode; in 1854 he had become a lecturer at the Government School of Mines and a naturalist with the Geological Survey, while his rôle as Darwin's supporter and publicist is too well known to need emphasising. He had been junior Secretary from 1872 to 1887 (Stokes was then senior Secretary), apparently very successfully.[49] Although Stokes was earlier rumoured to have thought him not to be Presidential material,[50] his stature was, by this time, considerable and he stood for a large group of biological and forward-looking scientists and educationalists. With Darwin clearly unavailable (although he had served on the Council in the early 1850s, he had long since retired to quiet and convenient invalidism at Downe) Huxley was a natural choice, fit in every way except for his somewhat belligerent character and his widely proclaimed agnosticism, still anathema to the majority of late Victorian scientists.

But once he was appointed to serve as interim President it was natural that the Council should choose him again when it met in October 1883.

Huxley himself was by no means anxious for the post; as he wrote to Michael Foster, his successor as junior Secretary,[51]

> The worst of it is that I see myself gravitating towards the Presidency *en permanence*, that is to say, for the ordinary period. And that is what I by no means desired. ——— has been at me (as a sort of deputation, he told me), from a lot of the younger men, to stand.

(It is worth noting the phrase 'the ordinary period'; only a dozen years after the completion of Sabine's long term a limitation had become accepted custom.) It is evident that Huxley felt that if he refused the Society might fall into worse hands than his own, or at least was unlikely to fall into better, for he told his friend William Flower (F.R.S. 1864, director of the Natural History Museum)[52] that although in June he had said that he

> ought not, could not, and would not take the Presidency...such strong representations were made to me by some of the younger men about the dangers of the situation, that at the last moment I almost changed my mind.

He stated firmly that the Council was, of course, free to choose its own candidate, but he seems to have realised that if he made himself available he was likely to be elected, and if he did not worse might befall, for he added

> I will not, if I can help it, allow the chair of the Royal Society to become the apanage of rich men, or have the noble old Society exploited by enterprising commercial gents who make their profit out of the application of science.

Yet he dreaded the strain involved for, as he told his daughter,[53] 'There was more work connected with the Secretaryship – but there is more trouble and responsibility and distraction in the Presidency.' In October, although Huxley had made it clear to Foster and others that he gave the Council a free hand, he was duly chosen, apparently without difficulty. In the event he was not destined to have much influence on the Society's affairs; within a year his health began to give way, and although he was re-elected in 1884 he spent the winter abroad and in 1885 resigned.[54]

The professional scientists 1885–1900

There seems to have been little obvious debate about the choice of Stokes to succeed Huxley. George Gabriel Stokes, six years older than Huxley, had been elected F.R.S. in the same year (1851); a Cambridge mathematician, Lucasian Professor since 1849, he had produced a large volume of work on applied mathematics and mathematical physics, the most original of which appeared before he became Secretary of the Royal Society in 1854. Although Huxley's name was, and is, better known to the general public than that of Stokes, most historians of science would regard the contributions of Stokes to have been more original than those of Huxley, who is now best remembered for his generous, enthusiastic and successful promotion of Darwinian theories of evolution. Stokes and Huxley were very different not only in their scientific outlook, but in their personal views. Stokes spent all his life in Cambridge, although he undertook administrative and teaching posts in London and elsewhere; Huxley, whose medical education had been at the Charing Cross Hospital in London, hardly stirred from London in the years after his voyage on the *Rattlesnake* except for holidays to recoup his health. Stokes was a convinced Anglican, active in reconciling science and religion; Huxley an agnostic who worked to separate the two. When Stokes in 1887 became M.P. for Cambridge University, Huxley, although he could say

> there is no one of whom I have a higher opinion as a man of
> science – no one whom I should be more glad to serve under, and
> to support year after year in the Chair of the Society

yet wrote a leader for *Nature* which, while expressing pleasure that Stokes had been elected, disapproved of his standing for Parliament on the grounds that he would sacrifice his independence to party (so that it was a waste of talent), and would make voting within the Society difficult, since by no means all the Fellows were, like Stokes, strongly Conservative.[55] Most Fellows seem to have agreed with Huxley that Stokes was an excellent President, even those who, also like Huxley again, disagreed with his political and theological views, while those in sympathy with his outlook naturally supported him fully. So no real dissension developed, and Stokes had a peaceable term as President. His greatest shock seems to have been that experienced in 1887 when, as he reported to the Council on 7 July, the Queen received the Society's loyal address on the fiftieth anniversary of her accession in company with the representatives of many other bodies, whereas the Society's officers had, by custom, been granted direct access

Fig. 19. William Spottiswoode (1825–83), F.R.S. 1853, Treasurer
1870–78, P.R.S. 1878–83.
By G. F. Watts

Fig. 20. Sir George Gabriel Stokes, Bart (1819–1903), F.R.S. 1851,
Secretary 1854–85, P.R.S. 1885–90.
Photograph by Barraud

to the throne. But a letter of protest to the Home Secretary resolved the matter which the Society had taken as a serious derogation of its privilege as a royal body.

Evidently there was some feeling that after forty years the 1847 statutes were out of date, for on several occasions during Stokes's Presidency the Council raised questions and constituted committees. One of these was a Publications Committee which in 1886 proposed to the Council (18 November) not only that papers printed for future inclusion in the *Proceedings* be distributed in advance of publication but that the President, with or without any consultation, might authorise any paper for publication or might refer papers on his own authority. In the end it was decided to give the President power to refer any paper to a suitable expert of his choice without consulting the Publications Committee, but not to give him the right to authorise publication. In 1888 more thorough consideration was given to possible statute changes. A radical proposal was the creation of a new class of Fellows to be known as Corresponding Members; the proposal was discussed by the Council on 5 July and although it was later withdrawn it was not without strong support in some quarters. The case for and against is well revealed in a letter from Huxley to the Treasurer, Evans (24 July, MM 9, no. 13); he wrote,

> I have carefully considered your draft statutes and I sincerely trust that they will be approved by the Council & then the Society –
>
> We shall look very foolish if, after all the talk, nothing is done to bring us into closer relation with the Colonial & American confréres [*sic*] – ...
>
> On all grounds I am extremely loth to go against Hookers judgment – but I cannot agree with him in this matter –
>
> Your scheme gives the officers no more influence than they have in the case of the Foreign Membership – and there, so far as my experience goes it is by no means paramount.
>
> I have not the slightest doubt that the Corresponding Membership would be highly appreciated both in America and in the Colonies and would afford deserved recognition to comparatively young men not only there but in this continent.

In the end the only important result of all Evans's proposed changes was the decision to permit the holding of an ordinary meeting after the annual election of Fellows; this both meant that more Fellows attended and that there was more opportunity for the reading of papers. In the next two years there were further discussions of possible changes in the statutes, notably those concerning the limitation in the number of new Fellows each year, but as Stokes reported in his last Presidential address (1 December 1890) the majority of the Council believed that such limitation was highly desirable, while the fact that the number of candidates in each year was steadily increasing was an indication of the prestige now attached to the Fellowship as a consequence of this limitation.

On 20 March 1890 Stokes announced to the Council that he would not serve for another year; presumably he felt that twenty-five years' service as an officer of the Society was enough, especially as he was now past seventy years of age. The Council Minutes for 3 July 1890 record the choice of Sir William Thomson, soon (1892) to be made Lord Kelvin, to succeed him. Thomson was a close friend of Stokes, whose work he had utilised in his earlier work on electricity; they had been contemporaries at Cambridge. But whereas Stokes was, scientifically speaking, a purely Cambridge mathematical physicist, Thomson, who had studied for six years at Glasgow before attending Cambridge, was deeply influenced by French mathematical physics, especially by the work of Fourier, and indeed had spent some months in Paris. In the 1850s he was to develop the ideas of Carnot (then little known in either France or England), working on various areas of thermodynamics at the same time that he began a long interest in the development of the Atlantic telegraph. In the 1880s, partly as a result of his involvement in the development of electrical lighting, he became a partner in the firm of Kelvin and White. Although so much in the forefront of the new physics, his geological ideas were less obviously progressive, for he was violently opposed to the current uniformitarianism, and he refused to accept Darwinian, or any other, evolutionary theory. From 1846 to 1899 he was professor of natural philosophy at Glasgow (much of this time being also a Fellow of Peterhouse, Cambridge), where he showed himself an inspiring teacher, although a poor lecturer. He had been elected F.R.S. in 1851, but had not served on the Council, although he had been active in the British Association, serving as its president at the Edinburgh meeting in 1871. His usual residence was Glasgow; however, his non-residence in London was perhaps offset by the advantage of the useful link between Scotland and England.

The first months of Kelvin's Presidency saw some debate on statute revisions, culminating in the new statutes of 1891: the changes were mostly minor ones, and included the provision of more flexibility in the conduct of meetings; no change was made in the number of candidates nominated yearly. Less pleasant was a personal attack upon Kelvin by Henry Wilde, an electrical engineer, F.R.S. 1886 (nearly twenty years after his first paper had appeared in *Phil. Trans.*). Wilde addressed a printed letter to the Council[56] dated 28 October 1891, entitled 'Declaration of Henry Wilde F.R.S. against Sir William Thomson P.R.S.', objecting to his nomination for a second year 'on the ground that, during the present year of his office as President of the Society he has not dealt faithfully in all things'. This was a farrago of attack ranging from Wilde's complaint that in his papers on magnetism Kelvin had used some of Wilde's work without due acknowledgement, had prevented the publication of some of Wilde's papers and rejected his conclusions, to an objection to a recently published advertisement for 'Sir William Thomson's Patent Indestructible Water Taps' because he was there specifically designated 'President of the Royal Society'. (By 1891 all papers were carefully refereed, so that it was most unlikely that Kelvin had anything to do with the rejection of Wilde's papers.) Wilde's letter was acknowledged by one of the Secretaries after its presentation to the Council on 5 November 1891, and there the matter dropped. No doubt the advertisement was more than a little embarrassing in the 1890s (many disapproved of making money out of 'pure' science) though Kelvin can be seen as in the earlier spirit of using natural knowledge for the benefit and use of mankind in devising superior, longer lasting taps which did not quickly begin to drip. A similar, perhaps more embarrassing occasion arose two years later, when, at the direction of the president and censors of the Royal College of Physicians their registrar wrote to

> call the attention of the President and Council of the Royal Society to the conduct of a Fellow of the Society, namely, to the manner in which he had been asking certain physicians for paid 'opinions' concerning a case in the Law Courts.[57]

After the Treasurer had volunteered to confer privately with the Fellow in question, a committee drafted a memorandum suggesting a code of conduct, which the Council duly adopted, and the matter was dropped without further publicity. An analogous problem was the use in advertisements of the designation 'F.R.S.' by those not members of the Society, a recurring point of difficulty over the years, once or twice the cause of apology by innocent offenders not themselves responsible for the designation.

In the summer of 1893 a major review of the Society's policy was instituted with the appointment of a Procedure Committee, chiefly apparently at the instigation of Michael Foster, to consider the conduct of meetings and the publication of papers.[58] Both these topics were of perennial concern: the Society was always aware of the problem of interesting Fellows in specialised papers as well as the necessity for improving the standard (and speed) of its publications. The development of specialised societies had led from time to time to discussions of the best rôle for the Royal Society in changing conditions, and now the Physical Society, of which Rücker, a member of the Committee, was currently president, had asked the Royal Society for a discussion of the best way to publish physical papers. Proposals ranged from suggestions (especially by J. N. Lockyer, F.R.S. 1869) that the Royal Society should virtually relinquish interest in subjects covered by specialised societies, through suggestions for affiliation with the societies (notably by Foster and Rücker), but in the end moderation prevailed, and a proposal by Glazebrook was adopted. This suggested that some ordinary meetings should be designated as being on single topics (as today); that not all papers should be read at length; that a number of the *Proceedings* should be issued as a result of each meeting; and that Sectional Committees should be set up to advise on papers and meetings.[59] These recommendations were finally implemented in 1896, having been extensively and carefully discussed by the Council during February and June 1895. The Sectional Committees have continued in increasing number ever since.

In his Anniversary address on 30 November 1892 Kelvin adverted to the contents of the *Philosophical Transactions* and *Proceedings*, and to their standard, and in 1894 he outlined the Procedure Committee's discussions about the conduct of meetings and the publication of papers, without being specific, while in 1895 he spoke of the financial necessity of limiting publication while trying to keep up with the supply, and described the Council's efforts to obtain some financial support.[60] It was left to his successor to announce the outcome of all these matters on 30 November 1896. Meanwhile rumours flew, and there were many attacks on the Society's administration, especially in *The Times*. Each year the report of the Anniversary Meeting and the Anniversary dinner gave occasion for editorial and epistolary comment. In 1892 this was fairly favourable, though the leader quoted a correspondent as objecting to the award of the Copley Medal to the German pathologist Virchow and as regretting that the Society sometimes appeared to stand less for science than for 'social advantages or personal friendships'. In view of subsequent criticisms it seems probable that the objection to Virchow does not reflect xenophobia, but a physical scientist's complaint at an award to a medical man.

In 1893 *The Times* published (2 December) a long diatribe by an anonymous correspondent who appears to have known more of the discussions of the Procedure Committee than had been made officially public: he claimed that there was far too much biology in the Royal Society (the first

meeting of 1893 had been devoted to a biological subject in accordance with
the proposed changes in the design of meetings); that Kelvin was dis-
satisfied with this state of things but was unable to correct it because he lived
too far away; that it had been proposed to confine the Royal Society to
branches of science without specialised societies of their own, which would
mean certain branches of biology; that the Council was too feeble; that there
were too many men from Trinity College, Cambridge, on the Council; and
that too much money had been spent on solar physics – hardly balanced
criticism, but certainly reflecting the views of a number of Fellows
imperfectly aware of what was being proposed. The *Times* leader for 1894
(1 December), while praising some things (including the Marquis of
Salisbury's speech at dinner), thought that there had been far too much
emphasis on biology, instancing the fact that Edward Frankland (F.R.S.
1853), as a chemist, had only just received the Copley Medal (Stokes had
been Copley Medallist the year previously), presumably again an objection
to the 1892 award to Virchow (which, be it noted, a modern would regard
as perfectly appropriate; while Hooker and Huxley had successively
received the awards in 1887 and 1888 there had not otherwise been an
excessive leaning towards biology in such awards). Whereas in 1894 the
Society had been rebuked for not having rewarded merit (in the person of
Frankland, then awarded the Copley Medal) sooner – he was then sixty-
nine – in 1896 it was rebuked for not having given a Royal Medal to a
'veteran' scientist whose name had been raised (J. H. Gladstone, F.R.S.
1853) but to a young man who could afford to wait, namely C. V. Boys
(F.R.S. 1888), at this time forty-one. The other Royal Medal went to
Geikie. There was other captious criticism; as Lister, now P.R.S., wrote
to Rayleigh, just retired as Secretary after eleven years in office:[61]

> You have no doubt seen the malicious attack in to-days *Times*
> about the R.S. The statements regarding the award of the Rumford
> Medals [to Lenard and Röntgen] and the second Royal Medal are
> of course quite false, and equally so is the insinuation that you
> declined to remain in the Council because you disapproved of its
> ways. It is felt by Rücker [Rayleigh's successor] and others as well
> as myself that a few words from you in the *Times* would do a very
> great deal of good.

In fact Rayleigh had already written his letter, explaining that his resignation
signified nothing except his desire to escape from London (which led to his
refusal to accept the Presidency in 1900, although he was finally to accept
it in 1905).

There is a certain irony in an attack on the Society for prejudice towards biology during the Presidency of Kelvin, a man so firmly on the physical side; the only justification could be a belief that the President's rôle was a mere sinecure, the power lying with the Secretaries, especially with Michael Foster (physiologist, F.R.S. 1872) who had been in office since 1888, only four years more than Rayleigh in 1896. The Council certainly backed Kelvin, as appears from a letter from him to Rücker, then on the Council, in February 1895, which showed his awareness of the desirability of a change:[62]

> Another thing I wanted to speak to you about, if I had had the opportunity at our last Council meeting, is the next election to the Presidency. I shall have had my five years, and it will then be time to choose my successor. It has been a great pleasure to me to be President, but for many reasons I feel that however kindly the Council might feel in respect to re-electing me a sixth time (as indeed has been suggested to me as an idea that might possibly be entertained) it would be in all respects better that another should be elected. I should be glad, therefore, if members of the Council would be considering in good time the question of who is to be elected

as indeed they soon did.

The choice fell on Sir Joseph Lister (F.R.S. 1860), now sixty-eight, a baronet soon to be created Baron Lister, the second surgeon to be P.R.S., who, although representing biology and medicine, had such a high reputation both personally and scientifically that there was no question that the Society would be regarded as favouring the B side in his appointment. On the contrary, when he had been urged to accept the Foreign Secretaryship in 1893, his friend T. L. Brunton (F.R.S. 1874, a distinguished physician) wrote[63] that he could help 'to heal...the breach that was forming and beginning to widen between the chemists and physicists on the one hand and the biologists, especially the physicians, on the other', while *The Times*, in the midst of its attack on the Society's over-emphasis on biology, approved the appointment only complaining because it had been ten years since Lister had been on the Council. What took place in 1895 was described by Lister in a letter to a friend with whom he had discussed the question:[64]

> When you left I had, as it were half consented, or rather...I had endeavoured to escape on the score of my utter aversion to being pitted against Evans [the Treasurer]. Michael Foster afterwards wrote & told me that there was such an almost universal feeling in

Fig. 21. William Thomson, Baron Kelvin of Largs, O.M. (1824–1907),
F.R.S. 1851, P.R.S. 1890–95.
Photograph by J. Stewart of Largs

Fig. 22. John William Strutt, 3rd Baron Rayleigh, O.M. (1842–1919),
F.R.S. 1873, Secretary 1885–96, P.R.S. 1905–08.
By Sir George Reid, 1903

the Council against Evans being President...that there was no
question whatever of my running in competition with him & that,
at the informal meeting of the Council which Lord Kelvin called to
talk the matter over 'they had no difficulty in coming to the
conclusion' that I ought to be President. I could wish most heartily
that it had been otherwise.

Lister was to be a quietly successful President, and in retrospect an
eminently proper and natural choice: he was personally acceptable to
everyone, it was time for a biological President, he was scientifically
eminent, he had leisure (his practice was now not great, he had retired from
his chair at King's College Hospital, and since the death of his wife in 1893
he had no home ties). There is now no record as to whether Lister was
the only choice – Sir William Flower may also have been considered[65] – but
there was almost certainly no one available on the biological side so truly
distinguished as Lister, even though he had to wait until 1902 to receive
a Copley Medal. He himself always thought that he was too much the
medical man, with too little knowledge of pure science, but the medical
profession was gratified. *The Lancet* writing of the appointment as 'fair and
judicious', noting though that the days when medicine had been closely
connected with the Royal Society because it was an easy introduction to
science were long past, yet medicine, like other sciences, should properly
be represented, and adding[66]

> of the impartiality [of the Society in respect to the various sciences]
> we may feel assured, for now, at any rate, virtue reigns supreme in
> Burlington House. And that is why there is not a savant in the
> world...who does not consider that to have been elected a Fellow
> of this body is an honour...

a supreme tribute to the respect in which Lister was held, if, inevitably,
a slight exaggeration, as *The Times* was to demonstrate a year later. But the
public attitude was on the whole closer to *The Lancet* than to *The Times*.

During Lister's Presidency there were a number of administrative
developments. In 1896 he could in his Presidential address announce plans
both for increasing the interest of meetings 'by giving greater freedom in
the conduct of them, and by enlarging the opportunities for discussion; and,
at the same time, more rapid judgement as to the value of communications
made to the Society', through the use of the Sectional Committees. He also
announced modification of the statutes and in particular the creation of the
Yearbook, which was to be published as soon as possible after the Anniversary
Meeting. From now on the *Yearbook* contained not only the calendar for

each session, the statutes and other information, such as the list of the Council and committees, but a detailed report of the Council for the preceding year and, during Lister's time, the Presidential address as well (it is now published in *Proceedings* A and B). From thence onwards, however, it was to be made plain to all that the major share of the Society's day to day work was the responsibility of the Council.

The Treasurers

The rôle of the Treasurer in the Society's affairs had throughout the century frequently been an active and powerful one, but not always a clearly defined one. Davies Gilbert, Treasurer 1819–27, had, during Banks's last year in office and during Davy's illness, been very powerful, and his holding of the office had helped him succeed to the Presidency. Although J. W. Lubbock, who served from 1830 to 1835, did not aspire to the Presidency, he played a very powerful part in running the affairs of the Society during the first five years of Sussex's Presidency, and was to do so again under Northampton from 1838 to 1845. George Rennie, who succeeded him and served under both Northampton and Rosse, does not seem to have made any notable impact on the Society, but after his resignation in 1850 the Treasurer stood out as a very important official, not only responsible for the increasingly complex financial affairs of the Society, but also acting as Senior Vice-President, *de facto* although not *de jure*. Sabine prepared the way for his Presidency by serving as Treasurer from 1850 to 1861. He was succeeded by W. A. Miller, F.R.S. 1845 when appointed professor of chemistry at King's College, London; he was an early worker in the realm of stellar chemistry and chemical spectroscopy, and a thoroughly professional scientist; his qualifications as a Treasurer perhaps derived in part from his post as assayer to the Bank of England and the Mint.[67] After nine years in office Miller died suddenly in late September 1870 at the age of fifty-three. It was his death, near the end of Sabine's ninth year of office as President, which precipitated the crisis which had at least something to do with Sabine's resignation, although Sabine denied this, claiming that it was because he was about to resign that he wished a trusted friend and colleague as his Treasurer and deputy. The facts themselves are not in doubt: Sabine wanted Gassiot as Treasurer, but the Council outvoted him and chose Spottiswoode who was to serve as Treasurer for eight years before being in turn chosen President in 1878, the last Treasurer to be nominated as President, as it happens.

Sir John Evans (F.R.S. 1864) was then appointed Treasurer. Evans was

Fig. 23. Joseph, Baron Lister, O.M. (1827–1912), F.R.S. 1860, Foreign
Secretary 1893–95, P.R.S. 1895–1900.
Photograph by Elliot & Fry

Fig. 24. Sir Michael Foster K.C.B. (1836–1907), F.R.S. 1872,
Secretary 1881–1903.
Photograph by Maull & Fox

a man whose background and interests were by this time unusual for a Fellow, although perfectly normal at the beginning of the century. He had no university education and could be classed as a wealthy amateur: he was a partner in a successful paper-making firm, who had become a specialist on questions of water-supply, on fossils and early stone implements, and was equally active in the Geological Society, the Numismatic Society and the Royal Society. He served successfully as Treasurer for twenty years, retiring in 1898 at the age of seventy-five; it is not clear whether it was his age or his mainly archaeological interests which militated against his possible nomination as President in 1895, when Lister was chosen; he was to be president of the British Association at its Toronto meeting in 1897. Evans was succeeded by Sir Alfred Kempe (F.R.S. 1881, mathematician and barrister) who was to serve for twenty-one years.

The details of the Treasurer's increasingly complex duties, as revealed in the Council Minutes, show that the financial affairs of the Society required an officer of diligence and financial acumen: land values were, on the whole, increasing (for example the Acton Estate held by the Society was only enclosed in the middle of the century, and almost immediately a portion was purchased for the use of the railways); there were new funds to be administered: the Gassiot Fund (established in 1871), for assisting magnetic and meteorological research (see Chapter 7); the Scientific Relief Fund, initiated by Gassiot in 1859, increased by donations then and subsequently, to assist scientists and their families in need (without respect to their membership in the Society); the Publication Fund established in 1878; and others, including monies provided by the Government yearly in the shape of the Government Grant and Fund (see Chapter 6).

The Secretaries and their assistants

With the President's rôle less dominant, and the Treasurers involved in ever heavier financial affairs, the duties of the Secretaries became greater, and more time consuming. From 1837 onwards there was always one Secretary representing the biological sciences and another the physical sciences, although they were not then so denominated (as the custom now is) but differentiated according to length of service as junior and senior Secretaries. In spite of the complaints at the length of Roget's term of office (1827–48), long-serving Secretaries were to be the rule rather than the exception, although by the end of the century a ten-year term was more usual than a twenty-year one. Thomas Bell, elected in opposition to Grove in 1848, retired in 1853 after only five years in office, upon his election as president of the Linnean Society, to be succeeded by William Sharpey, professor of anatomy and physiology at University College London since 1836, who had been nominated by Gassiot.[68] He had served several times on the Council since his election in 1839, and although not regarded as a great scientist was a decidedly influential teacher who was to prove an exceptionally hard-working Secretary, serving until 1872 when, like Roget, he retired at the age of seventy. His colleague on the physical side after Christie's resignation in 1854 (also at the age of seventy) was G. G. Stokes, who resigned in 1885 after thirty-one years in office to become President (he was then sixty-six). Stokes was, if possible, the most hard-working of all nineteenth-century Secretaries. As Foster, his colleague for many years, was to put it in 1884[69]

> Correspondence, etc. ought not and certainly will not in the future be as great as Stokes has made it. It has been painful to see how his energy has been wasted in this way.

Among many other administrative improvements Stokes introduced the use of the typewriter, about 1878 (capital letters only to begin with!), noting the advantages of being able to make clear copies of correspondence.

No other Secretary served as long as Stokes, although to his very great credit this long term of office seems to have provoked no overt criticism. As biological colleagues for Stokes there were first Sharpey; then T. H. Huxley, a very conscientious Secretary who served for only nine years (1872–81), retiring mainly out of the ill health which became manifest after his assuming the Presidency in 1883; and Michael Foster, known like Sharpey as a great teacher of physiology, mainly at Cambridge, who came closest to emulating Stokes by serving for twenty-two years, 1881 to 1903, retiring at the age of sixty-seven. He was partnered by Lord Rayleigh (F.R.S. 1873, before succeeding to the title) who served for eleven years (1885–96) until he tired of having to be so much in London, and A. W. Rücker (F.R.S. 1884), past president of the Physical Society, professor of physics at South Kensington, who resigned in 1901 upon appointment as principal of London University.

The post of Foreign Secretary, not a very onerous one, was twice held by Captain William Henry Smyth, R.N. (F.R.S. 1826), astronomer and geographer, first 1837–39, then 1850–56; between these two occasions J. F. Daniell (F.R.S. 1814, professor of chemistry, King's College, London) served 1839–45 and the ubiquitous Sabine 1845–50. In 1856 W. H. Miller succeeded Smyth, to be followed in 1873 by A. W. Williamson; in 1889 A. Geikie, geologist, was appointed, followed successively by Lister in 1893, Frankland in 1895, and another chemist, T. E. Thorpe, in 1899. The

Fig. 25..W. A. Miller (1817–70), F.R.S. 1845, Treasurer 1861–70.
Photograph by Maull & Co.

Fig. 26. Sir John Evans K.C.B. (1823–1908), F.R.S. 1864, Treasurer
1878–98.
By A. S. Cope, 1900

four-year term then became official in 1889; during his tenure the Foreign Secretary was *ex officio* on the Council.

Since the late seventeenth century the Secretaries had been assisted by someone not a Fellow variously called clerk, librarian, housekeeper or assistant secretary, which last became the official title in the mid-nineteenth century. The names of Stephen Lee and J. D. Roberton have already been mentioned. In 1843 Roberton was succeeded by C. R. Weld, a barrister with some literary pretensions. He served the Society well, especially in compiling and publishing his *History of the Royal Society* (1848), long a standard source particularly valuable for its printing of documents not accessible to those outside the Society. In 1844 Weld acquired as assistant (called initially 'sub-librarian') an odd young man, of little formal education but with a great yearning for a literary life, Walter White, whose *Journal* has often been referred to in the above account. When Weld was asked to resign in 1861, unfortunately under a moral cloud,[70] White was given the post, although it was very much sought after; he was a faithful and hard-working housekeeper and Assistant Secretary who clearly enjoyed his post, which he held until retirement in 1885. Although he was an eccentric, his eccentricities clearly created no difficulties with his scientific superiors.

As the existence of the post of clerk, an assistant to the Assistant Secretary, suggests, Weld had already made his post an administrative one. White's assistant, Wheatley, was to specialise in library affairs, although not officially the librarian, serving until 1879 when from a long list of applicants, Herbert Rix, a graduate of London University with a little experience of schoolmastering, was appointed clerk. He later succeeded White and served as Assistant Secretary from 1885 to 1896, when R. W. F. Harrison was appointed; he had served as secretary to the Society's Transit of Venus Committee and as assistant secretary to the City and Guilds Institute. By now the Assistant Secretaryship of the Society was a post of some responsibility; as Foster had written in 1884, urging Rayleigh to accept the post of Secretary considering that it was not so onerous as Stokes had made it, 'Mr. Rix is a very competent person, and can be entrusted with much more than he now has, and the Council I think will distinctly approve of this kind of work [routine correspondence] being taken off the secretaries', and the files show that White had begun to act as an executive secretary soon after his appointment. The library, editorial responsibilities, and much else devolved gradually upon the minor officials, with the Secretaries exercising 'knowledge and judgement' as Foster put it.

Papers, medals and candidates

On the whole, the precautions taken in refereeing papers, the existence of competent Assistant Secretaries and clerks to be normally present in the Society's rooms, and the publicity given within and without the Society to the decisions taken by the President and Council made things run smoothly. There were inevitably still complaints over the slowness of acceptance of papers (most usually caused by dilatory referees – some able men were regarded as improper referees because they were too slow in reporting on papers in spite of constant reminders), or the rejection of papers, when the unlucky author might sometimes accuse officers or referees of unfair bias. So in 1853 there was a long complaint by Lt. F. Higginson, R.N., who believed that he had seen the fall of a thunderbolt and had collected the resultant stones, thereby establishing the cause of magnetism; he was most indignant when the senior Secretary (Christie) told him that his conclusions were unacceptable, because he believed both in his own observations and that he deserved the award which he mistakenly thought had been offered for such a discovery.[71] The Secretaries tried to be patient in such cases, as they were when authors demanded rewards for 'discovering' methods of squaring the circle or 'inventing' perpetual motion, but as W. H. Miller groaned in 1870

> Perpetual Motion and the squaring of the circle is nearly too much, in one tract, for the most case hardened Foreign Secretary. The only novelty in the balance is the extreme clumsiness of the application of the principle on which it is constructed,

but he added 'I have answered his letter as well as that of the crazy "Psicologist"'.[72] In 1874 C. Piazzi Smyth actually resigned his Fellowship because a paper of his had been rejected, and then further complained because he thought the Secretaries had not sufficiently publicised his argument.[73]

A more unfortunate case – unfortunate because it appeared to strengthen the case against the biological Secretary, or rather to add fuel to the flames – occurred in 1895, when F. W. Pavy (F.R.S. 1863), a distinguished chemical pathologist, wrote to protest at Foster's handling of a physiological paper submitted during the summer. The author, D. N. Paton (F.R.S. 1914), had sent in a paper which had been refereed and approved for publication before it had been read, which, strictly speaking, was contrary to the statutes, as Pavy said, although it was quite normal for the officers to do this during the summer recess, and indeed they were always given blanket permission to deal with routine business at the last summer meeting of the Council. Pavy disagreed with Paton's espousal of Claude Bernard's theory of the glycogenic function of the liver (although he himself had as a young man studied with Bernard) and was particularly angry because while Paton's paper was published, his own in refutation was 'archived' and read only in abstract. After discussion with the President and Council, Foster wrote a public letter of apology, admitting that the officers had been at fault, as they now realised, 'in taking steps with regard to the publication of Dr. Paton's letter before it was read', and promising that the statutes would be amended (which was hardly Pavy's intention).[74] They were to be changed by the re-creation of the Sectional Committees in 1896, so that, ironically, what had been regarded as instruments of unjust partiality in the late 1840s became symbols of justice fifty years later. Total elimination of such grievances was of course impossible; in 1898 E. R. Lankester (F.R.S. 1875) was moved to declare that he considered that 'it would be advisable not to insert the referees' names in the minute book' and to erase all names previously entered.[75] The criticisms of the Society's affairs in *The Times* and elsewhere had emphasised the need for extreme confidentiality.

A related problem was the award of medals, for the criticisms associated with the 1847 reform continued from time to time, as the attacks in *The Times* in the 1890s demonstrated. At the very first Council meeting after Rosse's election (13 December 1849) a committee was formed to review the procedure for the Royal Medals, composed of Brodie (a Vice-President), Roget, Horner, Grove and Wheatstone and the officers, a fairly radical group except for the perhaps untactful inclusion of Roget, the ex-Secretary who had been at the storm-centre of the 1845 award. The committee, after canvassing prominent Fellows, agreed that both publication in *Phil. Trans.* and the previous advance notice of subject were unnecessary limitations, and suggested as new regulations[76]

> That the Royal Medals in each year should be awarded for the two
> most important contributions to the advancement of Natural
> Knowledge published originally in Her Majesty's dominions within
> a period of not more than ten years, and not less than one year of
> the date of the award,

regulations given royal approval three years later. There was now no restriction as to subject, although the Council at the suggestion of W. A. Miller (not yet Treasurer) resolved that it would be 'desirable' to give one Medal in 'each of the two great divisions of Natural Knowledge' (i.e. A and B), a desideratum generally kept in mind in the second half of the century.

All scientists now

From this time on, also, reports on the scientific work for which the medals were awarded were announced at the Anniversary Meeting and published, so that the reasons for the award might be thoroughly understood. So sensitive was the Society to the need for complete unanimity that in 1853 John Tyndall (F.R.S. 1852) tactfully declined the medal which he had been voted because he discovered that some of the Council had had second thoughts, and some non-Council Fellows had disagreed with the findings published in the award-winning paper,[77] leaving Darwin the only Royal Medallist in that year.

Election procedures as established in the 1847 statutes largely remained in force throughout the century. A minor modification made in 1853 was the resolution of Council that 'the Fellow whose name is first signed from personal knowledge, be requested to furnish...copies of such books or publications as may be referred to in the certificate', thus making for an even more formal and serious scrutiny of candidates. The choosing of Foreign Members was also rationalised: first a list of all those previously elected was drawn up, with nationality and field of specialisation; the achievements of those proposed in each year were to be described in considerable detail, so that all might appraise their importance. The election of the members of the category of Privileged Fellows was also reconsidered from time to time. In 1865 foreign princes regarded as royalty by the Queen had been added to princes of the blood royal, peers of the realm, privy councillors and foreign sovereigns and their sons as among those entitled to instant election (and supernumerary to the list of fifteen annual candidates); in 1873, after the failure of an attempt to assimilate all these to the procedure appropriate to ordinary Fellows, instant election was limited to princes of the blood royal and privy councillors, a regulation to last until 1902. Thus after 1873 no peers were elected unless their scientific merit was as great as that expected of commoners or unless they were privy councillors – which in fact meant that most extraordinary elections were of politicians.[78] So far had attitudes changed since the beginning of the century, when to have noblemen in the Royal Society was thought the only way in which its standing and influence in Government and Court circles could be maintained. Society had itself changed in this respect, although slowly; the Royal Society had here, as elsewhere, been in the vanguard of change, where once it had fought a rearguard action for the status quo.

5

The encouragement of science

When in his Presidential address at the Anniversary Meeting of 1857 Wrottesley told the Fellows there assembled that the Council had particularly turned its attention during the preceding year to the question of how best to improve the position of science in Great Britain, he reported that its conclusions came under the headings of education, the encouragement of discovery and research and the dissemination of an effective understanding of the importance of science to the community at large. In the event, it was the second of these aims that was most readily and fully achieved. Some methods of doing this already existed in the traditional award of honours. More novel – first attained only in the period after 1830 – was the award of monetary grants to individuals to cover the cost of their private research in whole or in part. Other methods were the improved distribution of information about new discoveries and the state of any particular science, and the direct involvement in specific subjects, most notably in the ancient, transparently useful but still unsatisfactorily scientific subject of meteorology.

Honours and awards

Most of the nineteenth-century ways of encouraging the pursuit of science were personal, and none more so than the award of some distinction to men whose scientific attainments and achievements were recognised as being substantially above the average. Nowadays, election to the Royal Society is itself such an award; as previous chapters have shown, this was not at all the case at the beginning of the nineteenth century, and hardly the case until after the 1847 reforms, when limitation of the number of Fellows elected in each year effected a change from an atmosphere in which choice was purely personal, as for a club, to one in which choice was, in principle, purely on scientific merit. And the limitation of the number of Fellows elected in any one year to fifteen made this principle virtually a fact, for any transgression would necessarily be glaringly obvious. Successful candidates were conscious of this, as they had not been earlier; there is all the difference in the world between Davies Gilbert's delight as a very young man without any scientific achievement at being patronised by Banks and so elected a Fellow, and Huxley's view that in 1851 at the age of twenty-six – only a couple of years older than Gilbert had been but with much solid, published work to his credit – he had been supremely lucky to be elected at his first candidature as one of fifteen (two of these were Stokes and William Thomson). No age of course possesses a monopoly of good judgement; not every group of fifteen was star-studded, which is not surprising when one considers how easy it is to magnify the importance of the contemporary who, in retrospect, is only second rate; but the first-class mind is less likely to be overlooked, and in the lists of candidates who were elected there is a very high proportion of those whose names the historian would expect to see. Even though the list of Fellows elected in 1900 may not impress the casual observer more than the list for 1800 – as it happens, there are no superstars in either – yet even the casual observer knows that those elected in 1900 will have their names in the *Catalogue of Scientific Papers*, and that those elected in 1800 will probably not do so. Indubitably, scientific achievement rather than mere interest was required for election in the second half of the nineteenth century, as it had not been earlier.

It might be wondered whether in straining after excellence pure science had ousted applied, and whether there were fewer technical men elected, proportionately, in 1900 than in 1800. It seems not. In the late eighteenth century distinguished and original instrument makers – Ramsden, the Dollonds, Troughton – were certainly elected, as were industrialists like Boulton and Smithson Tennant. The practice was to continue for the newer professions: civil engineers, like the two Brunels and four Rennies and later Rankine; industrial chemists, like Abel, Matthey and Mond; occasional mechanical engineers, like William Fairbairn who received a Royal Medal in 1860 ten years after election, or Charles Parsons; metallurgists, like Sorby who received a Royal Medal in 1874 and Bessemer; electrical engineers, like Wilde, Siemens, Swan and Ayrton; not to mention a fair number of army men in the technical services, from Kater and Sabine to Abney, noted for his work in photography; and naval men mostly at first associated with exploration, later, like Beaufort and Fitzroy, associated with hydrography. It should be further remarked that when the Society's advice was sought, as it so often was, it was mainly on practical problems; the scientific Fellows were never averse to serving on such committees but clearly believed fully in the Society's historic commitment to the application of science for the good of mankind. Eminence in applied science was therefore recognised as the equal of eminence in pure science; all scientific eminence was a qualification, provided it was indeed true eminence, and the Fellowship was its reward. It is noteworthy that although Huxley had given as one of his reasons for accepting the Presidency that he did not wish to see the office go to a 'commercial gent' who made money out of applying science to enriching ends, Kelvin was elected President in 1890 even if not without opposition.

If election to the Fellowship was the reward of scientific achievement, election to the Presidency was the supreme accolade, the acknowledgement of genuine scientific eminence. This was clearly so from about 1860 in contrast to the position in the first half of the century. Earlier it had been deemed necessary to choose a man of eminence and position to assist the prestige of the Society as well as that of science. In the latter part of the century scientific eminence was a sufficient criterion and as P.R.S. a distinguished scientist needed no other mark of prestige (although in fact he usually had some title, at least a knighthood, conferred upon him). The President of the Royal Society had become an acknowledged leader of science in Britain, so that the position not only was one of power but also a recognition of scientific, rather than, as earlier, of worldly, success.

Both less and more prestigious than the Fellowship itself was the receipt of one of the Society's medals – less prestigious because it might be conferred on a non-Fellow, more prestigious because given to fewer men. There were at the beginning of the century only two medals: the Copley and the Rumford. The addition of the Royal Medals, two in each year, brought the number to either three or four a year (the Rumford being awarded only every two years). While others were added towards the end of the century the first three carried the greatest prestige.

The oldest of these, the Copley Medal, has been since 1736 awarded annually, initially to the author of the most important scientific discovery or contribution to science by experiment or otherwise; in 1831 the regulation still prevailing was laid down, namely that it be awarded 'to the living author of such philosophical research, either published or communicated to the Society, as may appear to the Council to be deserving of that honour'.[1] It was always hoped that the existence of such a prize open to competition would encourage the pursuit of scientific knowledge by able scientists. The Rumford Medal (founded by Count Rumford in 1800) was rather in the nature of a pure reward, although equally intended to encourage discovery and improvement, in this case in the field of heat or light, if possible tending 'to promote the good of mankind'. The Royal Medals (1825) have already been discussed (in Chapter 2): originally there was an intention to make them directly competitive, like a prize competition of a French academy, but rapidly their terms were closely modelled on those of the Copley Medal, which, however, remained the senior honour. As has appeared above, the existence of both the Copley Medal and the Royal Medals did indeed act as a stimulus to scientists, who competed eagerly for an honour which was always coveted, although the competition did not by any means always result in that calmly disinterested engagement in the pursuit of natural knowledge which ideally, it was thought, the study of nature produced in its best practitioners. Later nineteenth-century medals – the Davy Medal of 1877 for chemistry, the Darwin Medal of 1890 for 'work of acknowledged distinction... in the field in which Charles Darwin himself laboured', and the Buchanan Medal of 1897 for 'distinguished service to Hygienic science or Practice' – are all in the nature of recognition of good work rather than promotion of it, and the practice which often obtained in the nineteenth century whereby individuals felt that particular papers submitted by them to the Society 'deserved' a particular medal no longer is possible, especially as even the Copley and Royal Medals are now more often awarded for a body of work than for a single contribution. Only those who have received them are in a position to state categorically whether their existence was in any true sense an incentive, whether the Royal Society's medals did and do truly 'promote' science or not. But it is hoped that the intention is a fact, and that it does so.

Money for research

Since 1828 the Royal Society has been in a position to encourage scientific research, not merely by the grant of distinction and honour for such research successfully undertaken, but by providing funds for the purchase of instruments and materials. It was Wollaston's bequest which created the Donation Fund, to which others, notably Davies Gilbert, Sir Francis Ronalds and Wheatstone, made sizeable contributions; later, in 1879, most of the money given by T. J. Phillips Jodrell in 1876 was added to this (as was to be the 1914 donation of Sir John Dewrance in 1954). It was, as noted above, at first little used (apparently only nine times before 1850, although a few, a very few, applications were rejected), even after a committee was appointed to consider its use in 1842; but after the establishment of the Government Grant in 1849, which familiarised scientists with the concept of the availability of research funds, it rapidly became more sought after. In the 1850s a good deal of money from Wollaston's fund went to the Kew Committee, mainly for meteorological and magnetic instruments; after that date, the grants (on average three a year) were often used as a substitute for a Government Grant, either when no more money was available there or when the objects or applicants were not suitable for a Government Grant (the Donation Fund was not limited as to a country or object, and was sometimes used to assist publication). In the last twenty years of the century the number of grants increased, as the amount of money available increased. The success of the Donation Fund may be partly judged by the creation of other similar funds, some with more specific objects: the Handley Fund for physical science, proposed at the death of E. H. Handley (never a Fellow) in 1840, though not received until 1876; the Gassiot Fund (1871) for magnetic and meteorological research, now paid to the Air Ministry in support of the Meteorological Office; the Joule Fund (1890) for research in physical science; and the Gunning Fund (1891) for research in physical and biological science. All these are now smaller than the Donation Fund, which has increased in the past 150 years. It seems a pity that the genuine innovation of Wollaston's bequest was never recognised by the bestowal of his name upon the fund created by it, but, admittedly, failure to re-christen it is consonant with his personal modesty.

An important contribution of Wollaston's bequest was its influence upon the outlook of those who as officers and Council members of the Society in 1849–50 were called upon to respond to the extraordinary proposal of the Government of the day (in the person of Lord John Russell) of a grant from the Treasury of £1000 to be used to promote science (see Chapter 6). Rosse as President was in favour of payment to assist experiments rather than as reward for success in research (the alternatives suggested) and after rapid and grateful acceptance of the proposal the Council on 13 December 1849 appointed a committee to decide how to handle what became known as the Government Grant. This committee was composed of the officers and four Council members, namely Owen, Murchison, W. A. Miller and Sir Frederick Pollock (F.R.S. 1816 for mathematical papers but now Lord Chief Baron; the desirability of having a legal expert to assist in setting up rules of procedure is obvious). The committee took as serious a view of the matter as the Council had done: £1000 was a very large sum and the Society had never previously been called upon to act for the Government as a distributor of money, although it had often advised the Government to undertake ventures requiring the outlay of considerable amounts of money. It is worth noting that in the committee's report, presented to the Council

on 7 March 1850,[2] it was recorded that all the members of the committee
had attended all its meetings. The committee's recommendations were as
follows: that,

> First and chiefly, the grant be awarded in aid of private
> individual scientific investigation.
>
> Secondly, in aid of the calculation and scientific reduction of
> masses of accumulated observations.
>
> Thirdly, in aid of astronomical, meteorological and other
> observations, which may be assisted by the purchase and
> employment of new instruments.
>
> Fourthly, and subordinately to the purposes above named, in aid
> of such other scientific objects as may, from time to time, appear to
> be of sufficient interest, although not coming under any of the
> foregoing heads.[3]

Although impressed by the responsibility the committee was by no means
overawed, for the report remarks drily 'that the application of considerable
sums in furtherance of scientific objects is not unprecedented in this
country, having formed an important feature in the objects of the British
Association for the Advancement of Science'. But the very fact that this
had been the case led some to anticipate problems: witness Whewell[4] (one
of those appointed to the first Government Grant Committee), who told
Murchison that

> I shall be very glad if plans can be devised by which the
> Government boon may be made a benefit to Science; but I think
> you will agree with me that to engineer it so as to produce this
> effect is not a very easy matter. We know the great difficulty which
> we had in managing undertakings supported by the money of the
> British Association; by no means mainly from the difficulty of
> procuring the money, but much more from the difficulty of
> avoiding the appearance and the reality of waste, caprice, partiality
> and jobbing. Some persons, I find, doubt whether the old practice
> of applying the screw of opinion in the scientific world to
> Government on each special occasion was not better than this
> perennial stream of bounty. But as the boon has been offered, an
> offer very creditable indeed to the spirit of the Government and the
> nation, it is very fit that we should try if it cannot be worked out to
> some good purpose, and I shall hope to attend the meeting on
> Thursday, and to hear what suggestions can be offered for such a
> purpose. I might perhaps make some myself, but I am not quite

sure that I like the responsibility of handling, or directing the handling of parliament money.

Clearly one problem was to ensure that the grants were made to deserving individuals, and impartially; equally important was it to ensure that the awarding of such grants should be seen to be unprejudiced and impartial. Hence the insistence of the planning committee that the 'committee of recommendations [be] of a very large and open character'. It was, wisely, proposed that the Government Grant Committee consist of the Council and of an equal number of Fellows not on the Council but 'conversant with the business of the Royal Society, or with one or more of the leading departments of science, or officially connected with the principal scientific bodies of the kingdom', each such member to serve for three years; and this was done. In 1859 it was decided that in addition to the Council and twenty more Fellows, the presidents of the Linnean, Geological, Royal Astronomical and Chemical Societies should be *ex officio* on the Committee; after 1862 this was an annual appointment, more convenient in view of continually changing circumstances (election of new officers of the various societies, changing composition of the Council, desire of members of the Committee to apply themselves for grants). On the whole things seem to have worked well, although not without requiring much care and thought; Rosse wrote in 1853 on the verge of retirement[5]

> The Government Grant originated in my time, & I am now endeavouring to derive some means of making the whole working of it as much a matter of business as possible.

Rosse was not at all worried about the expenditure of the Grant, although or perhaps because at this time not all the money granted each year was being spent. Rather he was concerned with the administration of individual grants, for which a special small committee was appointed in each case. In fact there was on the whole little difficulty; overseeing the use of the grants created no friction and the grants were expended for the purposes for which they were given. Hence the grants not only advanced science by permitting the recipients to purchase equipment they might otherwise have been unable to afford, but were, thanks to administration by the committees, clearly conforming to the terms on which they were allocated.

That this was seen to be the case was of great importance; it was essential that the Society be seen to retain none of the money for its own ends, and to make it perfectly clear that the Government Grant was in no sense a grant to the Royal Society, but a grant to science through the Society; that is, the funds were to be administered by the Society but wholly passed on to individual scientists, none at all being retained by the Society. It was also essential that publicity be given to the grants, for a wide dissemination throughout the scientific world, so that the Royal Society could be seen as being impartial in its distribution of the monies granted. The Society's old enemy, *The Lancet*,[6] while regarding the Grant with 'satisfaction' as an indication of the Government's 'wish to acknowledge the services rendered by Science to the State', inevitably doubted whether the Royal Society could be expected to apply it wisely seeing that 'Some of the Fellows elected to the Council have no scientific claims whatever', a statement sufficiently difficult to justify when the Council included J. C. Adams, Brodie, Darwin, Grove, Horner, Mantell, W. A. Miller, Murchison, Owen and Wheatstone; possibly Pollock's mathematical achievements had been forgotten, and exception could be taken to Sir William Reid known as a military man, who had however been elected F.R.S. for his work on hurricanes. The Royal Society's genuine success in administering the Grant was such that five years later even *The Lancet* was prepared to concede that it had been fair and impartial in its choice of recipients.

For some twenty-five years after its inception the Government Grant was administered by the Government Grant Committee which normally met in February and reported on its activities to the Council as soon as possible thereafter. Grants were given for all possible subjects, with every effort being taken to ensure the maximum distribution over the whole range of science. Amounts varied from £15 to, rarely, over £200; more usually they were in the range of £50 to £100. All equipment purchased out of the grants was to be the property of the Royal Society to be held after use 'for the benefit of the scientific public' as Wrottesley described it in his Presidential address, while in 1869 it was decided that the same rule should apply in the case of drawings. But a 'lending library' of instruments was no longer needed and a projected 'museum of instruments' never materialised; some suitable equipment went in the end to Kew Observatory, but most remained in use by the recipient of the grant. The grants were cautiously made and, as far as can be determined over a century after the event, in a reasonably judicious manner. Inevitably more grants went to Fellows than to non-Fellows (especially as more Fellows applied), and as is likely to happen in such situations there were many renewals. But there was never any sign of blatant favouritism and many of the applications refused were plainly unsuitable, being too expensive, too vague, too open-ended, or too obviously of doubtful value. When, in spite of the regulations, grants were requested for publication, the Government Grant Committee often turned instead to the Donation Fund. Initially the whole grant was allotted at one time, and worthy late-comers had to be referred to the Donation Fund, but later the Committee members understood the advisability of retaining some money in hand so that interim grants could be awarded at any time of year.

Then, in the spring of 1876, just after the Committee was forced to report that it lacked sufficient funds to support all the worthy requests made to it,[7] the Government, influenced by the 1875 Devonshire Commission's report on scientific education and the advancement of science which spoke very favourably of the existence of the Government Grant (see Chapter 6), offered for a five-year period to replace the annual grant of £1000 by one of £5000 to be given not only for equipment but for personal support; it was to be administered by the Science and Art Department, to whom all instruments should revert.[8] After consideration by the Council Hooker as P.R.S. wrote with an alternative suggestion, accepted by the Government, that the original Grant of £1000 be made and administered as before, but that the additional sum of £4000, to be called the Government Fund, be kept separate and separately administered. In the autumn it was determined by the Council that, in accordance with the Government's proposal, the presidents of the chief scientific societies, specifically including those in Scotland and Ireland, should be added *ex officio* to the membership of its Committee, and after the Anniversary Meeting the form of the new Committee was established: it was to have four sub-committees, initially A mathematics, physics and astronomy; B biology; C chemistry and D general purposes; sub-committee D was to draw up procedural plans and the existence of the new fund was to be extensively advertised.[9] On the whole this worked well and continued with minor revisions when a new Grant was made in 1882 on the expiration of the five-year period, and the situation remained the same through the remainder of the century. The Government Grant became one of the obviously overt ways in which the Society encouraged science, especially when, late in the century, the existence of the grants and the method of applying for them was advertised in scientific journals so that those outside the Society could become fully aware of their existence.

The dissemination of information

As the Society had long been aware, it was not enough to confine itself to encouraging and informing its own members; truly to promote science it was necessary that the results of research should be made available to a wider public, both at home and abroad. Thus it had long been generous in sending the *Philosophical Transactions* to other academies, societies and genuinely interested individuals, and after the *Proceedings* came into being these were distributed even more widely. Throughout the latter part of the nineteenth century the position was constantly reviewed (usually by the Library Committee) with regard to the free distribution and to the exchange of the Society's various publications and those published under its surveillance. The Society also undertook in mid-century to act as in some sense an agent, or rather distribution centre, for the reception of duty-free postal packets of books and periodicals sent from abroad: a valuable aid, although a decided onus on the employees (usually the assistant clerk or Assistant Secretary's assistant) who dealt with the matter. The Royal Society's involvement was made clear in March of 1852, when Rosse reported a discussion which he had had with a member of the Treasury Department about the 'admission, free of duty, of scientific works sent as presents to persons in this country'. The Treasury demanded a list of individuals 'likely to receive such presents', and such a list was drawn up by a committee who enlisted the help of other scientific societies; this list was to include all those with any pretensions to being professional scientists, who subscribed to foreign journals or who corresponded with some regularity with Continental colleagues.[10]

Thus the Royal Society for some years assisted scientists in the British Isles to obtain copies of scientific works published abroad, a useful, if minor, service to the world of British science. Far more important, because of its wider scope in every sense of the word, was the creation of the *Catalogue of Scientific Papers*, which as ultimately published was a listing of papers in the principal journals of the world on 'all branches of Natural Knowledge for the promotion of which the Royal Society was instituted'.[11] Appropriately, since in the twentieth century it was to become an international responsibility, the origins of the undertaking were themselves an example of international cooperation and understanding, with a touch of national patriotism thrown in. It is too well known to be discussed here at length, but, briefly, the origin lay in the belief of an American physicist, Joseph Henry, that the physical sciences stood in need of the greatest possible amount of international cooperation. In 1855, almost ten years after becoming secretary (the principal administrator) of the newly formed Smithsonian Institution in Washington, D.C., Henry sent a suggestion, which was discussed at the British Association meeting at Glasgow, for a 'catalogue of Philosophical Memoirs' to be drawn up by an international committee. The British Association Committee then constituted (consisting of Arthur Cayley, F.R.S. 1852, mathematician, [Robert] Grant, astronomer, and Stokes) was favourably inclined towards such a catalogue to be drawn up for all papers on mathematical and physical science from 1800, as a serial index. But nothing more was done until, on 5 March 1857, Sabine, still very active in British Association affairs, brought the report to the notice of the Royal Society's Council, requesting on the behalf of the Association the Society's cooperation. A committee was formed, consisting of the British Association committee together with Professors De Morgan, Graham and W. H. Miller, to consider the matter fully; on 18 June they reported to the Council that after careful consideration they thought that the index was feasible and desirable, and should consist initially of the papers published in the *Philosophical Transactions* and the memoirs and journals of other scientific societies published since 1800.[12] After this, it was referred to the Library Committee (which by this time included, besides the officers, George Busk, Cayley, Hooker and Wheatstone, so that the medical, mathematical, natural and physical sciences were amply represented) which reported to the Council on 7 January 1858 that it recommended making in the first instance a manuscript catalogue of all the periodicals in the Society's own library, including *all* the sciences, this to be done at the Society's expense, including the payment of 'one or more' assistants. At the next Council meeting the sum of £250 was voted for the purpose, and at the Anniversary Meeting that year Wrottesley as President could announce the inception of the *Catalogue*.

Work went on so well that by January 1861 two additional assistants were required; soon other institutions were solicited for titles, both at home, and, later, abroad. By June 1864 the *Catalogue* was so far advanced that the Library Committee could put forward proposals for its printing, hoping that the Government would finance it, a proposal which Sabine as President could announce as accepted at the Anniversary Meeting of 1864. This first volume, published in 1867, set the pattern, which remained unchanged in successive volumes. The first six volumes cover the period 1800–63; then in 1873 came the decision to publish further volumes for the decade since 1864 (published 1877) and after this volumes 9 through 11 covered the period to 1883, while in 1902 appeared a twelfth volume containing more items for the period 1800–83. In 1889 there was the alarming news that the Government would no longer produce it as a Stationery Office publication, but it was quickly discovered that the Cambridge University Press would undertake publication, as it did. Work continued steadily during the later years of the century, although on 30 June 1892 it was resolved by the Council that in future only titles 'found in the Libraries of the principal Scientific Societies in London' should be included, the Committees 'having power to omit such of them as appear to be unimportant', the whole indicating the magnitude of the task, now becoming almost too great, and requiring extra assistants. (Initially such assistants were boys, fresh from school; later women were employed instead.) At the same time the Council began to consider whether this vast task ought not to become truly international. This decision was postponed after Dr Ludwig Mond (F.R.S. 1891), a highly successful industrial chemist and manufacturer, made in 1892 a donation of £2000 which permitted the publication of volumes 10 and 11; he later gave additional sums towards an Author Catalogue (1884–1900). Subsequently three volumes of the Subject Catalogue were also published. For the period after 1900 the Society acted as publisher of what became an international undertaking.

Why, it may be asked, was the *Catalogue* undertaken, and why was it hailed as being such a valuable contribution to the promotion of science? It was a novel concept, for it must be remembered that there were few purely scientific journals in the modern sense when the *Catalogue* was conceived, beyond those which were the transactions of the societies which published them as a record of papers read at meetings. The one important exception was the *Philosophical Magazine* (or *Phil. Mag.*; on occasion, indeed, this published papers read at meetings of the Royal Society and elsewhere for it could ensure quicker publication than could most societies, the Royal included). Crookes's *Chemical News* was only founded in 1859 and Lockyer's *Nature* in 1869. There were throughout the nineteenth century few reviewing or abstracting journals and as the scientific world expanded it was difficult, indeed almost impossible, to keep abreast of published literature on a particular subject. All scientists tried to maintain a wide correspondence with colleagues, but even this could not ensure that important papers might not be overlooked; indeed the scientific world was accustomed to the fact that important papers were often missed and so the discoveries they announced had to be re-made, especially when the scientists concerned lived in different countries. After all, foreign journals were not then readily available in any country, even in capital cities, in spite of the system of exchange of journals usual between analogous learned societies. So for example although a Fellow of the Royal Society had potential access to a very wide range of foreign and domestic journals (provided that he visited the Society's Library) it was by no means the case that even he had access to every paper in his own field, and non-Fellows were inevitably at a much greater disadvantage; besides to look through the transactions of many foreign societies was time-consuming and often difficult. The *Catalogue of Scientific Papers* made it possible for almost any serious scientific worker to discover what work had been done in his field and by whom, and in many cases showed him how to gain access to it. The *Catalogue* was thus a monumentally effective undertaking, which was of immense utility to the scientific community at large. That it was in the later twentieth century to become equally useful to historians of nineteenth-century science was an unanticipated contribution to scholarship in that area which lies between science and the arts. The *Catalogue* only ceased to be essential when the ever greater increase in science made abstracting journals in each scientific field a necessity.

Meteorology and magnetism

By its very nature the Royal Society aimed to promote all science equally, and the birth throughout the nineteenth century of societies devoted to individual sciences tended to increase the importance of this. Equally such sciences as had no specialist societies might seem especially worthy of support. This last may in part explain why the Royal Society devoted a great deal of effort to the promotion of meteorology and magnetism; for although the Meteorological Society was founded in 1851, the Physical Society was not founded until 1873. Both meteorology and magnetism were sciences useful to the Government, of an international character and in need of central organisation to promote and develop world-wide observations.

Meteorology and its associated instruments – barometers, thermometers, hygrometers – had been of concern to the Society since its earliest days. In the eighteenth century the Society had prided itself on providing information to travellers on instruments and methods of employing them, while in November 1773 the Council had resolved that observations with 'barometer, thermometer, rain-gage, wind-gage, and hygrometer' should be made twice daily at the Society's house, the results to be published annually in *Phil. Trans.* This was faithfully done from 1774 to 1843. In 1822 a committee was established for examining the instruments used, and a new barometer was installed, showing how seriously this work was then taken.[13] Further activity was shown in the 1830s, when a Meteorological Committee was regularly re-appointed each year.

In the 1830s also meteorology took a new turn, and the Meteorological Committee faced new problems and took new decisions. One result was the discontinuance of regular meteorological observations in the Society's rooms; as the Joint Committee on Physics and Meteorology[14] noted, in June 1840,

> the meteorological observations as at present conducted are
> unavoidably unworthy of the official character which they bear, in

consequence of the imperfection of the locality and the multiplied
duties of the observer

and therefore recommended to the Council that it 'desist from causing
observations to be made under circumstances so unfavourable'. Suggestions
were made for alternative sites, thereby offending Airy who pointed out
indignantly that observations were currently being made under his supervision at Greenwich, and that he could readily continue, extend, and
ultimately print them, an offer promptly accepted, after which publication
in the *Phil. Trans.* ceased.[15]

Of much greater ultimate importance was the extension of the concept
of meteorology to include the study of the intensity of magnetism as
recorded throughout the world and the establishment of magnetic observatories. This was partly based upon the widely held belief that terrestrial
magnetism was essentially identical with, or at least closely connected with,
atmospheric phenomena, a belief held by such disparate scientists as
Humboldt, Arago, John Herschel, Sabine and Faraday. Gauss dissented,
holding that magnetism was a purely terrestrial phenomenon, but the
developing knowledge of the connection between electricity and magnetism
reinforced the majority view, as did Christie's discovery in 1831 that the
Aurora Borealis influenced the earth's magnetic field. However Gauss
agreed with Humboldt that many systematic observations were necessary
and both worked to promote interest throughout Europe, while Gauss's
theoretical work was immensely important in offering a substructure to the
observations being carried out. Among British scientists particularly
concerned with promoting the collection of magnetic data were Sabine,
Airy, John Herschel and Humphrey Lloyd (professor of natural and
experimental philosophy at Trinity College, Dublin), all of whom were
associated with both the Royal Society and the British Association, which
had begun to foster the subject at a very early stage; indeed Sabine's later
scientific life revolved almost entirely about the study of terrestrial
magnetism.

Hence the Royal Society was well prepared to react favourably to
Humboldt's famous letter of 23 April 1836, requesting cooperation in
establishing a chain of magnetic observatories and the despatch of an
antarctic expedition primarily for magnetic study,[16] while in the next year
the Society urged the East India Company to sponsor such work,
subsequently advising those entrusted with it, as well as dealing directly
with the Rajah of Travancore. Magnetic study – what has been somewhat
flamboyantly called the magnetic crusade[17] – was a continuing concern of
the Royal Society, while more particularly the interest of the British

Association, although it must be remembered that those active in promoting this work were members of both institutions. Thus from 1841 to 1871 the British Association sponsored and maintained Kew Observatory (adjacent to, but outside, Kew Gardens), after which date the Royal Society took it over.[18] A number of grants were made to support it from the Donation Fund in the 1850s and 1860s but the Society's continued administration of Kew would have been impossible without the generosity of Gassiot, who offered the Society funds to provide an income of £250 a year for the maintenance of a 'Central Magnetical and Physical Observatory'. This enabled continuance not only of magnetic studies but of other experimental physical work by both the staff and outside scientists, until in 1899 Kew was converted into the National Physical Laboratory, of which the Society was to have ultimate control, the President and officers being *ex officio* members of the governing body, a responsibility that ceased only after the First World War. (See below, Chapter 7. In fact, Kew had long since ceased to be a suitable site for magnetic observations, for it was increasingly surrounded by first railway and then electric railway lines.)

While in 1840 meteorology proper seemed to be submerged beneath the rising tide of magnetism, it was to re-emerge soon as a subject in its own right, exemplified in the creation of the Meteorological Office in 1855. Initially, as appears from the Board of Trade's request to the Royal Society for advice,[19] the aim was analogous to that of the 'magnetic crusade': the accumulation of data upon which 'great atmospheric laws' might be framed. In the event, a more practical immediate end was settled upon, the gathering of information which might help shipping, but still with the belief that data were to be accumulated before any 'laws' could be framed. The primary function of the Office was to be the collection of information about the strength and direction of winds, variation of the temperature of the air and the ocean, variations in barometric pressure, and the characteristics of ocean currents. This information was to be obtained particularly at sea, and although some might be obtained at seaside centres it was not then thought necessary or even advisable to establish a series of stations on land, particularly in view of the systematic observations regularly conducted at Greenwich and other observatories.

The Meteorological Office received a somewhat different direction as a result of the appointment, at the Royal Society's suggestion, of Captain (later Admiral) Robert Fitzroy, formerly Captain of the *Beagle* during Darwin's voyage and F.R.S. 1851 in consequence of the excellence of his hydrographical work. Fitzroy was not content to collect and analyse data; in spite of his formal title of Statist (statistician) he wished to utilise the data as he collected them, being all too conscious of the frequent loss of life at sea, and even more by shipwreck on the coast, as a result of unanticipated storms. (He had been particularly shocked by the disastrous wreck of the *Royal Charter* on Anglesey with great loss of life, a disaster which, he thought, could have been prevented by better meteorological knowledge.) He realised that from the reporting of storms at sea (made increasingly quickly as telegraphic communication improved) and knowledge of wind direction, it was possible to predict what storms might reach coastal areas and when, and he began a valuable service of storm warnings. These were much appreciated by seamen, although there were complaints that fishermen might be put off from going to sea by the prediction of a storm that never materialised. Fitzroy devised a crude but innovative symbolic display of weather cones; and he also devised a method of charting the weather, equally crude and equally innovative. Realising well the imperfection of his method he nevertheless believed in its value enough to publicise what he called 'forecasts' – a word he believed less indicative of certainty than any other – and those which he made available to the press could be seen daily in *The Times* from 1862 onwards, where they attracted favourable comment. Naturally his forecasts, even his storm warnings, were by no means perfectly reliable, but they were very interesting and on the whole helpful. Sadly, Fitzroy took as directed against him personally the views of those such as Francis Galton (F.R.S. 1860) who had a horror of anything that could be regarded as 'unscientific', that is, not derived as absolute laws based upon an immense quantity of data, and who believed that meteorology above all subjects required to be on a purely 'scientific' basis. Public controversy certainly played a part in Fitzroy's suicide in 1865, which in turn led to a formal investigation, a Parliamentary report and a total reorganisation of the collection of meteorological information.

As the *Report of a Committee to consider certain questions relating to the Meteorological Department of the Board of Trade* (1866) makes clear, the views that the Royal Society advanced were really those of Galton, who dominated the committee of three (a representative of the Royal Society, one from the Board of Trade and one from the Admiralty, but none from the Meteorological Department itself, although information was sought from Fitzroy's principal assistant). In Galton's view, Fitzroy's work had been wantonly, as it were wickedly, unscientific.[20] He was particularly shocked (and was to continue to be so) because no one had been rigorously thorough in the collection of data and, in particular, when storm warnings and forecasts had been made there was no explicit description of the reasoning on which they were based and no careful analysis of success or failure. The Royal Society, he insisted, could not associate itself with such a procedure; it represented *science* and could not sanction the publication and dissemination of any but 'scientifically' based information, that is, information derived directly and recognisably from detailed data. The collection and analysis ('digestion') of data was the thing needed, and all that he could approve. So in 1868 the Meteorological Department of the Board of Trade was replaced by the Meteorological Committee of the Royal Society with Galton still active on it. Daily forecasts of weather were discontinued, although weather charts were still made available to the public, and storm warnings were curtailed, although not totally abolished since public outcry demanded their continuance.

Data were collected as before, and the Meteorological Committee acted not as a Royal Society committee but as a government department, except in name. This anomalous position, as its head was to call it, was made public during the Devonshire Commission investigation of 1875. Finally, in 1877 Galton and other Fellows dramatically resigned *en bloc* declaring that in their view the Meteorological Office should 'as in other European countries, form a department of Government', as the report stated. After some negotiation the Government established the Meteorological Council to replace the Committee; its members, although mainly nominated by the Royal Society, were paid, and no longer reported directly to the Society. Even this responsibility was lessened by the replacement of this Council in 1905 by a Treasury Committee with two representatives nominated by the Society, an arrangement to be continued when in 1919 the Committee was converted into an advisory body to the new 'Met. Office' under the Air Ministry. Throughout the nineteenth century the Gassiot Committee and the Kew Observatory encouraged wider applications of meteorology, including collection of data on atmospheric electricity, seismology and geomagnetism.

No other subject received so much direct attention in this century as did meteorology. Royal Society committees were set up on seemingly innumerable subjects, both physical and biological, but these were mainly in response to outside pressure – from the Government, from individuals, from other societies, or on account of expeditions planned primarily for other reasons, in connection with which the Society's help was sought (see Chapter 8). The concern with meteorology was partly fortuitous, partly the result of long-standing interest within the Society and partly the result of equally widespread interest in various parts of the community.

Varied interests

By the last quarter of the nineteenth century there were other subjects with
whose promotion the Society was explicitly concerned, either because of
the degree of organisation required or because of the pressure of individual
Fellows. These were normally subjects requiring Government assistance.
Thus the Society's traditional association with Greenwich Observatory
makes it seem natural that it should have exerted pressure on the Government
to assist in organising expeditions to observe the transits of Venus in 1874
and 1882, as it had done in 1761. Solar eclipses demanded more attention,
as well as the aid of Government (to mount the necessary expeditions) and
the cooperation of other societies; this was particularly true in the 1870s
and 1880s, while in 1896 the Joint Permanent Eclipse Committee was
established (jointly with the Astronomical Society) to supersede *ad hoc*
committees for another fifty years. In 1883 the spectacular eruption of the
Krakatoa volcano produced the Krakatoa Committee whose report was
published five years later with dramatic coloured illustrations, a great and
important novelty. Solar physics (1884), stellar photography (1887), coral
reef boring (1895–1900), evolution (1894 at the instigation of Galton – really
concerned with biological statistics), dredging (W. B. Carpenter's survey of
marine life), exploration, all produced eponymous committees to advise and
assist individuals who sought the Society's help, usually in conjunction with
applications to the Government Grant Committee. Provided that the
subject was promising and the investigator competent almost any subject
might expect a welcome, with advice and assistance furnished freely in
accord with the Society's explicit aims.

One other aspect of the promotion of science, historically speaking, is the
question of the amount of recognition given to different branches of science.
This is not easy to appraise, and impossible to quantify. No new subject
seems to have been explicitly excluded in the bestowal of Government
Grants; if, therefore, a subject seems unrepresented it is impossible to
determine the reason, for so much depended on individual initiative. If
someone with a new idea failed to apply for a grant, nothing was done; even
if a man tried and failed one can conclude nothing definite, for then as now
grants were awarded not only on the basis of subject and individual
competence but on the basis of the feasibility of the project and the
likelihood that it would produce results. A more promising source for
judging the Society's view of the state of science of the day might be taken
to be that portion of the President's Anniversary address in which he
described the scientific achievements of the preceding year. But these were

always individual appraisals, and as the century advanced and specialisation increased were concerned mainly with subjects with which the President was familiar – A or B as the case might be.

In summary one can only say that without exception the Presidents of the second half of the nineteenth century were remarkably conscientious men who did their best to present the world of contemporary science fairly, aware of the fact that the Royal Society existed to promote science as a whole and that this meant the recognition of all scientific advances, on all fronts, however radical such advances might be. And although there were so often accusations of bias and of reactionary favouritism directed against the Council near the mid-century, the honours awarded to Darwin and his doctrine demonstrate the contrary. Consider the careers of Owen and Huxley. For many years Owen held great power within the Society, being often on the Council and influential in the affairs of the Physiology Committee; he rightly enough received a Royal Medal in 1846 and a Copley in 1851 for his work on palaeontology, which Darwin greatly respected, and this before he had reorganised the natural history departments of the British Museum and become the director of the new independent Museum. Yet Huxley, also the recipient of a Royal Medal for zoological research in 1852, was held in higher esteem than Owen for his espousal of Darwinian evolution, receiving the Copley Medal in 1888, and he served as President of the Society, an honour never conferred on Owen. Of the Presidents in the latter part of the century only Sabine and Airy were so elderly as to appear to belong to past science, and both of these kept well abreast of advancing ideas, in their special subjects at least, if not, in Sabine's case, biological science. Hooker maintained a studied impartiality as far as he could, and, like later Presidents, was still actively concerned with advancing his own subject, as Rosse had been at mid-century. Later Presidents represented a broad spectrum of science and were receptive to the newer trends in their colleagues' activities. They were thus fitted to represent science as a whole to the Society, and the Society as a whole to the scientific world at large, anxious that the Society should promote science in general and be seen to do so.

6

Relations with Government

As the first of the chartered societies, and chartered so early that it was truly 'royal' in the sense that the granter of the charter was the King, not his ministers, the Royal Society has occupied throughout the centuries a special place in the country in relation to Government. This has been a two-way relationship, involving privilege (ranging from direct access to the Crown, a jealously guarded right, to access to various government departments) and responsibility (to advise all parts of the Government when requested to do so).

Now everyone knows that, besides the charter, all that Charles II, titular founder, ever gave to his Royal Society was the title and a mace, and that he more than once derided the Fellows' addiction to seemingly useless experiment; yet even he regarded it as a potentially useful aid to Government, and was on occasion, if never often, prepared to ask its advice. Later sovereigns went further: so Queen Anne made the Society overseer of the Greenwich Observatory founded in the reign of Charles II, and began to associate the Society not only with the Observatory but with problems relating to the navy. First was the determination of longitude at sea, later in the eighteenth century other aspects of navigation; in the early nineteenth century came questions of the protection of wooden ships in tropical waters, and the management of compasses in iron ships. By the mid-nineteenth century investigation of problems at the request of the Government had become a commonplace, ranging from physics through chemistry to geology and from mathematics to medicine.

The other side of the coin was that the Royal Society itself felt free to approach the Government of the day directly, just as it had always had the right of direct access to the sovereign, and this spread to the development of more or less close relationships with various government departments, ranging from the Board of Trade to the Foreign Office, the Admiralty and the India Office and (at the end of the century) the London County Council, either to ask for help or to offer advice. While in 1850 it had seemed necessary that the President of the Royal Society should sit in the House of Lords, so that he might have access to influential members of the Government of the day and, as Rosse put it, give the Society 'political weight', by 1890 it was the Society which conferred upon its President the position of influence which made access to government sources smooth and easy. By 1900 the Society was approaching the twentieth-century attitude which has made it the official representative of science in the British Isles.

The Government Grant

The most innovative event in the relations between the Royal Society and Government was clearly the institution of the Government Grant in 1849, which instantly put the Royal Society in the position of being a major patron of science. The inspiration came from government sources, possibly influenced by a growing opinion in what would now be called informed circles that national prestige demanded that Britain should officially do more to foster science, combined with the beginning of the belief which culminated in the Devonshire Commission of 1875 that education must be reformed to provide a more scientific basis for English life and industry.[1] Although it was not always realised in the years that followed, the Royal Society's only direct contribution was the 1847 reform which made it a suitable recipient of the Government's plan, as set out by Lord John Russell (F.R.S. 1847 in the Privileged Class) who as First Lord of the Treasury offered Rosse, as P.R.S., a grant from the Treasury of £1000 to be used to promote science. Later members of the Government were inclined to believe that the impetus must have come from the Society, but that this was not so was demonstrated plainly by Russell's letter, which it is therefore worth quoting as far as it has survived.[2]

> As there are from time to time scientific discoveries and researches which cost money and assistance students of science can often but ill afford, I am induced to consult your Lordship as President of the Royal Society, on the following suggestion:–
> I propose that at the close of the year the President and Council should point out to the First Lord of the Treasury a limited number of persons to whom the grant of a reward, or of a sum to defray the cost of experiments, might be of essential service. The whole sum which I could recommend the Crown to grant in the

present year is 1,000 l., nor can I be certain that my successor
would follow the same course; but I should wish to learn whether,
in your lordship's opinion and that of your colleagues, the cause of
science would be promoted by such grants.

Rosse himself promptly replied favouring payment for experiments, rather
than rewards; the Council merely sent thanks

for the confidence which, on the part of Her Majesty's
Government, his Lordship has reposed in the Council and the
grateful sense the Council entertain of the liberal zeal for the
promotion of science, and the encouragement of scientific men,
which is exhibited in the offer thus made.

As noted above (Chapter 5) the Society itself had little difficulty in
adjusting to the administration of the Government Grant, novel though it
was. Things went smoothly for several years, the Grant being renewed
annually, although the Society had constantly to stress that it was only a
trustee for the Grant, keeping nothing for itself, even for administration.
Then on 8 November 1855 the Society and indeed the whole scientific
community was shocked when the Government Grant was withdrawn by
the Treasury, on the grounds that the money had hitherto been granted from
sums intended for charitable purposes, which were too limited to admit of
a continuation of any annual grant.[3] The Treasury letter, however,
inferentially suggested that the Society should apply for a direct Parliamentary grant, and this Wrottesley as President warmly welcomed, carefully
emphasising the point that the Grant was not in any way used by the
Society, but was regarded 'as a contribution on the part of the nation
towards the promotion of science generally in the United Kingdom'.[4] The
matter was debated in the House of Lords at the instigation of Lord
Brougham,[5] although it seems probable that Wrottesley's influence with the
Government was more important. By the autumn the matter had been
settled, the Treasury had agreed to recommend an annual vote of £1000
to be made by Parliament for this purpose and Wrottesley was able so to
inform the Council on 8 November 1855. The £1000 Grant continued for
the remainder of the century, for the Royal Society was careful to ensure
that the Government's generosity of 1876 (the offer of an extra £4000)
should be distinguished as the Government Fund, since both its purpose
and its source were different, and it was feared that if at any time the larger
grant ceased the lesser, original grant might have been lost if they were
amalgamated. (There was, in fact, never any real fear that this would
happen.)

The Government and science

It must be stressed as significant that it was to the Royal Society that the Government turned, even though it had not had as much experience as the British Association had had of distributing funds for research. However much attention other societies might attract from specialists, or the public, there was clearly no doubt but that the Royal Society held a special place in public life, being peculiarly responsible for acting as an intermediary between the Government and science, and the only possible intermediary between Government and individual scientists. At a time when the Science and Art Department was not yet in existence, being itself the product of the growth of public appreciation of the importance of science which produced successively the Government Grant and the Great Exhibition of 1851 (the immediate cause of its foundation) there existed no department or office in the Government which took any direct interest in fostering art or science, but only private organisations. This indeed was taken to be a characteristic feature of the British scene, one which distinguished British from Continental societal organisation.

It is a further distinguishing feature of English society in the nineteenth century that the Government could and did concern itself with the needs of the scientific community and that by mid-century legislators saw the need to foster science and assist individuals in their scientific research. That they did so in 1849 is sufficiently indicated by the institution of the Government Grant; and although changes in Government as a result of the downfall of ministries might threaten its existence, the temper of the times and the energy of the Society itself ensured its continuance. In 1856, by which time the Government Grant had been regularised and accepted as an annual event, Wrottesley as President could report in his Anniversary address that Parliament had been discussing how to improve 'the position of science or its cultivators', and the creation of a Board of Science was under discussion; if this did come into being it was possible that the Royal Society would need a special committee to work with it. (When the Department of Science and Art came into existence, its principal connection with the Society was, later, the responsibility for the Government Fund of £4000.) On the other hand, in 1861 Brodie, in his Anniversary address, could declare that he personally did not have any complaints at what the Government had done in regard to science, for the Royal Society had never been like a foreign Academy 'entirely subjected to the Government, without whose approbation the election of a new member is incomplete', and this he sturdily maintained to be one of its strengths. He saw the Society and the Government as having

'a mutual interchange of good offices', under which the Government
usually attended to the advice and requests of the Society and allowed its
annual Grant to be at the disposal of the Society, thereby avoiding any
excessive utilitarianism. As already noted, the Society lost no opportunity
of reminding the Government and the public at large that neither the
original Government Grant nor the enlarged Grant of 1876 was for the
benefit of the Society, being all distributed to individuals.[6] It was felt
important to maintain a balance in relations with Government, lest the
Royal Society's complete independence be jeopardised.

Not until the very end of the century was it the case that the Government
conferred any money directly on the Society. True, from time to time the
Government assisted the Society indirectly as by paying for the publication
of work initiated by the Society, most notably the *Catalogue of Scientific
Papers*. But as this was published by the Stationery Office almost as long
as it was the Royal Society's responsibility, no money passed through the
Society's hands. It was only in 1896 that the Government first gave to the
Society monies to be expended on its own publications in the form of a direct
Treasury grant.[7] The only other *direct* benefit the Society received was its
housing. Ever since its move to Somerset House in 1780 (in Banks's first
years) the Society had regarded it as normal that the Government should
provide it with a 'house', this to accord with the Society's view of its needs,
not the Government's merely. Thus it had sought from time to time more
room in Somerset House, and when it became apparent that it must move,
the Society firmly resisted the Government proposal that it should go to
South Kensington, demanding – and getting – accommodation nearer to the
centre of things, in Burlington House (Piccadilly) which was conveniently
close to Parliament and to the Athenaeum, to which many Fellows, as well
as many politicians, belonged. Similarly, as already noted, the Society
expected, demanded, and in the end always obtained, its ancient right of
direct access to the sovereign, its patron both nominally and (in respect of
the Royal Medals) *de facto*.

The Royal Society as advisory body

Government appeals for advice to the Royal Society continued unabated
after 1830, and took many forms. At the same time the Society itself
constantly appealed in turn to the Government on behalf of its individual
Fellows for assistance in carrying out proposed researches. Sometimes the
relationship was virtually symbiotic, in so far as both the Society and the
Government were concerned to promote the same causes. It is, for example,

a little difficult to decide whether in its capacity as a Visitor of Greenwich Observatory the Society was acting as an agent of Government or as an intermediary between the Astronomer Royal and the Government, both factors being involved. As already noted, however, the Society's rôle was modified by William IV, when the Board of Visitors was first enlarged by the addition of members of the Royal Astronomical Society, but *ex officio* participation in the Board of Visitors continued until the mid-twentieth century. (Nowadays the Society has representatives on an Advisory Committee of the Science Research Council, which has replaced the old Board of Visitors.) Astronomical interests continued to be promoted throughout the nineteenth century, especially by requests for government assistance, as will appear below.

Analogous to the Royal Society's relationship with Greenwich Observatory and with the Board of Longitude (see Chapter 2) was the rôle played by the Society in assisting and advising the Admiralty and the Board of Trade over the collection and utilisation of meteorological data. At first the Society acted in its normal advisory capacity; later, as noted above (Chapter 5) its interest and the wish of these government departments for detailed guidance resulted in a period of some ten years (until 1877) when the Society was *de facto* in charge of the government Meteorological Department, even while the Government itself refused to recognise its responsibilities, insisting that (although money to run the office was provided by the Government) it was a private interest of the Society. This was the closest the Society ever came to acting as a branch of Government, and it was careful to avoid such a situation during the remainder of the century.

The profound interest in magnetic observations also led to a curious interweaving of the interests of the Society and of government interest, which is difficult fully to unravel. There is no doubt whatsoever that the Royal Society was responsible for creating an active participation within the various government departments in the collection of magnetic data, but soon this participation seems to have taken on a life of its own, so that in turn these departments were to seek the aid and advice of the Society, as if it had not been the Society which was initially responsible for the whole far-flung network of magnetic observatories. The Admiralty began instructing its captains on the desirability of collecting magnetic as well as meteorological data, in almost equal detail, and to this end by 1837 was seeking the advice of the Society concerning the best manner of organising these readings, while by 1845 the East India Company was deeply engaged in collecting information through its officers and through the interests of the Rajah of Travancore (modern Kerala, in the southern tip of India), who was particularly helpful. In the years after 1858 there was an immense amount of business arising from the work of the Schlaginweit brothers, German naturalists in India whose enthusiasm was great and who posed a number of problems for the East India Company, which duly sought help in their solution from the Royal Society. All this was very dear to the heart of Sabine, Treasurer in the 1850s and President during the 1860s.

No other subject was as preoccupying as the related topics of meteorology and magnetism, partly because no other subject was of equal interest to both the Society and the Government. However, the Society constantly entered with zest into subjects raised by the Government and the committees appointed were always conscientious and generally anxious to be as helpful as possible. Thus for example from 1832 to 1838 the Excise Committee worked in response to a request from the Treasury: its task was to devise a simple and practicable form of hydrometer to be used by the excisemen in assessing the duty on spirits, wine and beer, and this it duly did, to the complete satisfaction of the Treasury. Then the interest already in existence for devising a standard pendulum, work undertaken by both Kater and the young Sabine, was turned to the problems of standard linear measures after the loss of the standard yard in the fire which destroyed the Parliament buildings, where it had been kept. The Society offered advice in the 1840s and in 1843 at the request of Airy, who as Astronomer Royal was deeply involved, agreed to allow the standard yard and pound held by the Society to be borrowed. (This was an unusual concession, for in the normal way no instrument or object was allowed to leave the Society's rooms.) Again in 1864 both the Treasury and the Ordnance Survey sought – and obtained – advice on the standard metre. Work was completed in 1853, when copies of the standard yard and pound were formally deposited with the Society.[8] In 1866 the Society was to offer advice in regard to the Great Trigonometrical Survey of India, while in the early twentieth century the Society coordinated work on the measurement of a geodetic arc throughout Africa.

The Society's involvement in projects for metrification (as it was then correctly called) in Great Britain was, like that of the Government, ambiguous. An International Decimal Association was set up after the 1855 (Paris) Exhibition, and Great Britain had representatives on the Association (apparently private individuals); the fourth meeting in 1860 was even held at Bradford (Yorks). There had been a memorial to the Chancellor of the Exchequer in 1859, urging the great savings to be achieved in educational effort by decimalisation. But the general temper of the times was firmly opposed to any mandatory introduction of metrification; the most that was achieved was a permissive Bill (the Metric Weights and Measures Act of

1863) legalising metric measures. The Royal Society took no official part in all this. When in 1867 a private individual requested permission to compare his own (glass) metre with the Royal Society's platinum standard his application, supported by the Council of the International Decimal Association, was refused,[9] although he merely wished to use his metre, as he said, 'to assist in familiarizing the Public with the Metric System by distributing copies of this the Standard Metre in England', the reason given being 'that the subject is likely to come under the consideration of a Government Commission'. It did so in 1868, when all the members of the commission were F.R.S.,[10] but although the commission's report was in favour of metrification, neither the Society nor the Government pursued the matter for some decades.

By the 1880s, however, the Society had become deeply interested in the international standardisation of units, and thus ultimately in metrification. From 1881 to 1883, at the request of the Department of Science and Art, the Society represented Britain at the International Commission on electrical units held at Paris. In 1884 the Society offered advice when the United States raised a question about the international acceptance of the Greenwich Meridian (the Americans wanted an American meridian). Finally, in 1887 the Society sent representatives to the International Metric Convention. In this last case the Society felt it incumbent upon itself to urge the Government to continue to be represented on this important committee; the Government of the day was not then much concerned, but the Society was able to insist on the importance of participation in the Convention's deliberations, even though Britain had still no intention of 'going metric'. In this the Society was so successful as to become identified in government circles with a firm grasp of the problem, so that in 1897 it was necessary to form a committee to deal with a query from the Board of Trade about the exact relations between metric and imperial units, the Astronomer Royal (W. H. M. Christie, F.R.S. 1881) having refused to take part in the discussion of the problem.

Some government requests for aid represented long-standing interests. Thus for many years the Society's favourable advice had been largely responsible for the government support (to the tune of over £7000) for Babbage's calculating machine, and it was no fault of the Society's that the support finally ceased: rather, as already noted, Babbage lost interest in his first design, and obstreperously failed to take active steps to complete the machine upon which so much work had been done. As late as 1855 Rosse, in his Presidential address, while regretting that no work had been done on the machine since the 1830s could note that he had tried to persuade the

Government to finance and organise work on it in 1852. But in 1854 the Government was more interested in a calculating machine devised by 'M. Scheutz' (George Schütz, a Swede) who had taken his initial idea from a report about Babbage's design, not known to him in any detail. The committee appointed to consider it (Stokes, W. H. Miller, Professor Robert Willis and Wheatstone) could not but report favourably, wisely as it turned out, for it was rapidly built and proved very serviceable in the Registrar-General's office.[11] Schütz was even suggested (by Cayley) for a Copley Medal in 1856, and although he did not receive one, the proposal says much about the Society's attitude to his success, and to the favour with which utility and government service were regarded.

Other government requests for assistance were more wide-ranging. Thus when in 1859 the Governor of New South Wales advocated the preparing of 'a work or works on the Zoology, Botany, and Geology of our Colonial Empire', the Society promptly established a committee to advise the Colonial Office on the desirability, feasibility and manner of achieving this end, of which they highly approved; the committee emphasised the importance of terrestrial physics (i.e. the favourite magnetism) as well. The Colonial Office, of course, had under its wing world-wide magnetic stations, and in 1862 asked the Society's opinion about the desirability of building a large telescope for the Melbourne Observatory, larger indeed than any in existence in the southern hemisphere (in fact, as the Society's letter pointed out, the Royal Society and the British Association had twelve years earlier urged the desirability of having a large telescope located in the southern hemisphere). In 1864 the Colonial Office approached the Society about the trigonometrical survey then in progress, and particularly about proposed pendulum experiments to be carried out by the superintendent, an army officer Lt. Col. Walker; this provoked a flood of letters from various Fellows consulted by the President, which were forwarded. Routine advice was still being given in 1866.

In 1865 the Society was advising the Admiralty about the problems inherent in 'the very great increase which has taken place in the employment of iron in the construction and equipment of ships, and the consequent augmentation of the embarrassment occasioned in their navigation by the action of the ship's magnetism on their compasses', as Sabine's official letter as P.R.S. to the president of the Board of Trade put it;[12] the matter had been raised at meetings earlier in the year. The Society's 'Memorandum' dealt with the effects on the compasses and the correction of the errors thence arising, something normally done satisfactorily in the navy, but not so successfully in the merchant marine. The reply was that the Board of

Trade saw the Society's points to be important and cogent, but impossible to carry out since the law did not permit coercion; but the provision of a manual or set of directions for the use of merchant seamen would, it seemed certain, be most desirable. The Society obtained the navy's directions to consider and subsequently urged their adoption to prevent the loss of life by wreck of passenger ships; the whole was published in the *Proceedings* (1866), but little came of the Society's very reasonable concern and endeavour to prevent disaster. (It seems that the Board of Trade was then more interested in the Society's opinion about a proposal to employ 'illumination by means of electric light' in the pearl fishery in Ceylon.)[13] The Admiralty was, generally, more prepared to consult the Society on its scientific and quasi-scientific problems, ranging from sending in 1853 a sample of some glass bottles found on the coast of Nova Zembla for examination (the committee found them to be merely Norwegian net floats)[14] to requests for advice about the application of copper sheathing and stripping to iron ships for protection against atmospheric electricity.[15]

As the century wore on the Society continued to supply assistance and advice to various government departments on a wide variety of problems. In 1879 four Fellows were nominated to assist the work of a Royal Commission investigating the cause and possible prevention of colliery explosions, as had been done earlier in the century,[16] while in 1895 the Home Office sought and received help in determining the cause and prevention of explosions of gas cylinders.[17] In 1888 the Board of Trade sought advice about the best material to be used for illumination of lighthouses[18] – again an investigation reminiscent of past work – and in 1896 it requested the formation of a committee to investigate the behaviour of steel rails under stress, for the railways found that they were losing strength with time and use.[19]

Biological problems

By no means all the committees working on Government-initiated problems were concerned with the physical sciences: by the last decade of the century the Society's help was being sought over a variety of biological and medical problems. Indeed in the last quarter of the century the Society itself had from time to time offered unsolicited advice on such problems.

In 1876 there was much agitation among physiologists on the subject of the anti-vivisection bill (the Cruelty to Animals Act) then being considered by Parliament. This sought severely to restrict experimentation on living animals, initially limiting it to experiments directly related to 'saving or prolonging human life, or alleviating human suffering' (not allowing it for the sole purpose of advancing knowledge without direct practical ends in view), and forbidding the use under any circumstances of dogs and cats.[20] As the Bill passed through the House of Lords it was opposed by various interested bodies, especially the medical profession and by a number of peers like Lord Rayleigh who understood the position of research physiologists. (Ironically, Lord Carnarvon, who was in charge of steering the Act through the House of Lords, had been elected F.R.S. in the previous year.) Physiology was at this time increasingly orientated towards animal experimentation, and it was precisely such experimental physiologists who were most likely to be Fellows of the Royal Society. The great difficulty was that public opinion was strongly in favour of the Act and it was all too easy for those promoting it to represent scientists as monsters unconcerned with animal pain in their selfish search for knowledge; besides, with no regulation it was clear that at least some unnecessary experimentation had been done, particularly in the teaching of students. The British Medical Association was the strongest organised opponent, but the Royal Society did act since scientific principles appeared to be involved. In mid-July the Council authorised Hooker as President to send a reasoned letter opposing the excessive restriction of the Act, pointing out that it was impossible for any scientist to predict which particular experiment might be *directly* applicable to human welfare, and arguing for a wider degree of latitude in the use of animal experimentation in physiological research. And on 22 July a deputation consisting of Hooker, Sir James Paget, Michael Foster (doubly concerned as physiologist and Secretary of the Royal Society) and Professor John Burdon-Sanderson (a noted physiologist, F.R.S. 1867 and strongly opposed to the Act) tried to present the physiologists' point of view to Carnarvon. The Bill was in the end considerably modified, experimentation on cats and dogs and a wider range of experimentation being allowed, although only under licence by inspectors (one of whom was Busk, a member of Huxley's *x* Club). The Royal Society thus played only a minor rôle in this liberalisation of the anti-vivisection bill; it openly intervened because scientific principles were clearly involved, since the original Bill would have meant the end of much physiological investigation in pursuit of knowledge, investigation which might or might not soon or even ultimately be useful or applicable to medical practice.

Connection with scientific principles remained a criterion for intervention in medical problems. Thus in 1880 Spottiswoode as President joined with medical colleagues in a deputation to Parliament on the subject of a proposed vaccination bill, feeling it proper to act (although the Council was not then sitting) because, as he put it at the Anniversary Meeting, 'the remedy proposed appeared to trench so closely upon the application at least of a scientific principle' that he thought it right to intervene.[21]

Later in the century there was a number of Royal Society committees called into being by the Government on biological and medical subjects. A notable instance occurred in 1890, when the Board of Trade sought the Society's help in investigating the problem of the detection of colour blindness. This was of concern to the railways, with the introduction of coloured lights for signals, and also to the merchant marine, in connection with buoys and with port and starboard lights on ships. In March 1890 a Colour Vision Committee was appointed which deliberated for two years before producing a report covering all aspects of the problem and making detailed recommendations for tests of future employees.[22] Meanwhile, in 1891 the London County Council's appeal for help resulted in the initiation of the Water Research Committee, which for the next five years was to investigate the twin problems of the detection of pathogens dangerous to human health and the treatment of water to ensure its purity.[23] This became an expensive piece of cooperative research, paid for half by the L.C.C., half by the Government Grant. The authority of the Royal Society in such matters was clearly growing greater if anything, as suggested by the report of the officers to the Council on 26 October 1899 that the Society had been requested to give as much publicity as possible to the memorandum of the Government of the Straits Settlements offering grants for research into beri-beri, a nutritional disease only identified a couple of years earlier.

In the last years of the century the Society's help was sought over a variety of purely medical problems. Thus in 1896 the Colonial Office asked for assistance in the investigation of the presence of the tsetse fly as a disease vector in Africa and, as a by-product, a study of the dissemination of rinderpest.[24] The Tsetse Fly Committee was prompt and thorough, securing the services of a competent scientist to go to Africa for field work and also gathering evidence from those familiar with tropical medicine. In 1897 the Treasury asked the President to inform the India Office of the existence of an anti-toxin for bubonic plague, then epidemic in India;[25] this is a curious instance of the Society's serving as a coordinating agent between government departments. In 1898 the Malaria Committee was to begin a five-year deliberation at the request of the Colonial Office, studying the character, distribution and importance of the disease in Africa, again including some field work.[26] No doubt this sudden increase in purely medical investigation was partly the result of the fact that so eminent a medical man as Lord Lister was President of the Society during the last five years of the century. Perhaps too there had been some influence of the

Devonshire Commission's finding (see below, pp. 177–78) that research in biological science had been relatively neglected in the first three-quarters of the century and starved of money and government promotion. If so, this neglect was being radically remedied.

Approaches to Government

There is another aspect to the Society's relations with Government, namely the fact that the Society's position and authority was such that it could successfully appeal to various government departments for assistance in initiating scientific investigations, especially on behalf of individuals who were desirous of undertaking research which required the aid of government bodies. The most obvious example is the case of the Society's influence in the endeavour to obtain magnetic observations from every region of the earth. To this end the Society devised rules and tables and advised on the use of instruments, so that magnetic readings could be taken by naval officers, by some merchant marine officers, by army officers in India and Egypt. Further it was not the Society but the Government, in one guise or another, which sponsored the establishment of magnetic observations in the colonies, and the India Office particularly continued actively the support of the observations begun earlier under the sponsorship of the East India Company.

In the seventeenth century the Royal Society had listened attentively to the efforts of John Wallis, Savilian Professor at Oxford, to construct a theory of tides based primarily on observations taken around the English coast, and had encouraged other men to describe tidal phenomena. In the 1830s Lubbock and Whewell both worked on the theory of tides, receiving each a Royal Medal (1834 and 1837 respectively). This work would have been impossible without the cooperation of the Admiralty, which in 1831–32 endorsed the collection of data about tides throughout the world, wherever naval stations were to be found or naval ships penetrated, while in 1852 the Admiralty provided ships for a tidal survey by a naval captain under the supervision of the Astronomer Royal, and paid for the publication of his memoir on the tides of the North Sea and English Channel.[27] While no doubt the navy might be expected to benefit ultimately, the primary reason for the collection of data was scientific.

There were other examples of research initiated by individual Fellows and sponsored by the Society of less obvious ultimate utility but which because of the Society's influence obtained assistance from the Admiralty or the War Office. A most striking example is to be seen in the development of marine zoology and oceanography, notably in the 1860s. Here the individuals concerned were W. B. Carpenter (F.R.S. 1844), physiologist and at this time registrar of London University, who had turned to the subject about 1860 through an earlier interest in microscopy, and Charles Wyville Thomson (F.R.S. 1869), successively professor of geology, zoology and botany at Queen's College Belfast, professor of botany at the Royal College of Science in Dublin and, from 1870, professor of natural history at Edinburgh, who had particularly interested himself in the marine biology of the sea depths. It was the inspection of specimens obtained by Thomson from a Swedish attempt at relatively deep-sea dredging which fired Carpenter in 1868 with the idea of seeking support from the Society in approaching the Admiralty for assistance. As he wrote,

> Such an exploration cannot be undertaken by private individuals, even when aided by grants from Scientific Societies. For dredging at great depth, a vessel of considerable size is requisite, with a trained crew, such as is only to be found in the Government service.[28]

This was in June; by August Carpenter and Thomson were at sea in H.M.S. *Lightning* (a paddle steamer) dredging in the North Atlantic, where they obtained good specimens at depths below 1800 feet and observed variations in the temperature of the sea such as had not been noted before. The next year the Admiralty provided H.M.S. *Porcupine* for cruising off Ireland and Shetland, where more specimens were obtained and work was begun on analysing the composition of seawater at varying depths. All this, with the Society's constant support, was to lead to their most important venture, the voyage of H.M.S. *Challenger* (1872–76), the results of which filled fifty volumes, published between 1880 and 1895 (see Chapter 8).

No doubt the example of Carpenter and Wyville Thomson encouraged others to ask for help for large-scale schemes which also required major assistance, often from the Admiralty, impossible for individuals to secure without intervention by the Society. Thus in 1883 a plan was put forward for making deep borings in the Nile Delta to ascertain its stratigraphy,[29] only a year after the uprising in Alexandria which resulted in the assumption of control over the Delta by the British army. Huxley was interested in the idea, and as President shortly afterwards not only assisted in the securing of a small grant from the Donation Fund but applied to the War Office for assistance. The work was carried out by army officers over the next five years under the supervision of the Delta Committee (all undeterred by the excitements of the siege of Khartoum and the campaign against the Mahdi in the Sudan) with some money from the Government Grant, for the work, which took the borings to 300 feet, was very expensive. A similar enterprise was sponsored in 1895 when the Coral Reef Committee was set up to supervise an investigation of the structure and composition (animal and mineral) of the coral reef of Funafuti (Ellice Islands), this time with the assistance of the Admiralty; work continued for the next five years.[30] It will be perceived that these successful researches would have been impossible without direct government assistance, which private individuals could never have obtained; they also generally benefited from monetary assistance from the Government Grant administered by the Society, and government aid for publication. Mention must also be made here of the numerous expeditions either initiated or sponsored by the Society which would have been impossible without government aid, and which were variously established to further geographical, physical, astronomical, geological and biological knowledge (see Chapter 8).

The Government and science

All this cooperation with the Government had not been received with universal approval, right and responsible as it appears to modern eyes. The views of Babbage and others in the early 1830s have been discussed above (Chapter 3), and Babbage felt in 1851 about the relation of science and Government as he had in 1831 – partly neglected because his calculating machine was not shown at the Great Exhibition, partly annoyed by the small rôle played by pure science in that event, which had been initiated by the Society of Arts, Manufactures and Commerce. (Significant, perhaps, of the truly improved relations between science and Government was the fact that the Royal Society was approached for advice in connection with the International Exhibition of 1862.) In 1847 an anonymous 'F.R.S.' had published a 100-page *Thoughts on the Degradation of Science in England*, insisting that no change had occurred since the earlier attacks of Babbage and others,[31] and that it was still the case that

> the sciences and arts of England are in a wretched state of
> depression, and that their decline is mainly owing to the ignorance
> and supineness of the Government...and to the indirect
> persecution of scientific and literary men, by their exclusion from
> all the honours of the State.

'F.R.S.', while attacking the attitude of the Government, believed firmly in the benefits of government support in all aspects of life, and his pamphlet is devoted to examples of inefficiency caused by the lack of such support and the exclusion of scientists from the running of almost everything, from the British Museum (then of course still including the natural history section) to the foreign service.

But soon such an attitude became impossible. The work of Wrottesley in the 1850s both before and during his Presidency, had been of great service in bringing the utility of scientific men to the possible notice of the Government, for patronage or employment or advice, and in 1856 he could triumphantly inform the Anniversary Meeting that Parliament was then discussing how to 'improve the position of science and its cultivators'. He obviously expected approval, yet not every proposal for closer association of the Society with the Government was cordially welcome: in 1849 Rosse, discussing at the Anniversary Meeting the success of the Government Grant, had felt compelled to note that some had even thought it might be 'injurious' to the Society to administer government funds. Although this view was soon dispelled (except among the few who for one reason or another – often merely pique – doubted that the Grant had been properly administered) there were still those who, like Brodie in 1861, could rejoice that the Society was and would remain independent of Government.

A somewhat different position emerged from the investigations of the Devonshire Commission, as it was usually called, more properly the Royal Commission on Scientific Instruction and the Advancement of Science.[32] When initiated in 1870 it mainly sought evidence on the teaching of science in England and Scotland, naturally mostly in the universities (Oxford and Cambridge were excluded); but it soon turned its attention to 'the Advancement of Science'. The Commission itself consisted of eleven members, of whom only three were not F.R.S. (and one of these, Bernhard Samuelson, was to become F.R.S. in 1881), and it included Sharpey, Huxley, W. A. Miller (replaced on his death by H. J. S. Smith, Savilian Professor of geometry) and Stokes, while those who gave evidence were mainly professors of science or army and navy officers in scientific positions (most in these categories being F.R.S.) together with what would now be called professional educators. They were expected to bring prepared evidence and they did. Many complained about the lack of government encouragement for their particular subject: so for example W. B. Carpenter (3 May 1872) claimed that biological science received no encouragement, and many physicists deplored the fact that there were no museums of the physical sciences comparable to the natural history museums then in existence (a peculiarly nineteenth-century point of view).

When Sabine appeared to give evidence the question naturally turned on the relations between the Royal Society and Government, for he was there as P.R.S. He was particularly asked whether 'as a rule' he found 'the Government willing to entertain favourably any recommendation from the Royal Society', to which he replied 'Remarkably so; they have always been so, and are certainly not less so at the present time than they were formerly', and he professed to be unable to remember any 'important' cases in which government assistance had not been forthcoming. A full discussion of the terms of the Government Grant then occurred, and Huxley gave Sabine the opportunity of stating categorically that the Royal Society received nothing for its labours of administration – a valuable reminder to the public, which had so often assumed the contrary. (Indeed, the Commission's inquiry also made it possible to emphasise the amount of other work undertaken by the Fellows on behalf of the Government without pay – notably in the case of the Meteorological Committee – and the Society's use of its own funds for the good of the scientific public – notably in preparing the *Catalogue of Scientific Papers*, on which over £3000 had already been expended.) The result of this testimony, bolstered by the knowledge of the Commission itself, was a favourable impression of the Royal Society's relations with Government, and a recommendation for an increase in the Government Grant, possibly with some of the money, at least, to be allocatable for personal remuneration, it having been carefully pointed out by several witnesses that so far all funds had been for apparatus and similar expenses incurred by the experimenter, and none for his time. This recommendation was to be implemented during the very next year in connection with the £4000 Government Fund administered by the Department of Science and Art, which was kept distinct from the £1000 Government Grant; the committee making the awards was able to work to different rules from those governing the original committee, and personal expenses became allowable.

Even more than the increase in government money allocated to science and scientists, various witnesses stressed the importance of having a definite body to advise the Government on scientific affairs. This idea had first been raised in 1856, when, partly through Wrottesley's influence, Parliament had debated the question of improved aid to science; it decided to propose the formation of a committee later in the year, and as a consequence the Council of the Royal Society decided (11 July) to offer its opinion to the committee which, it was hoped, would 'consider the question, whether any measures could be adopted by the Government or Parliament that would improve the position of Science or its cultivators in this country'. The Government

Grant Committee, chosen to discuss the matter, presented its report to the Council on 15 January 1857; this included as its final proposal the creation of a board to advise the Government on matters of science, this board to be created either by recognising the President and Council as the official government scientific advisers or by means of an entirely new body 'somewhat after the model of the old Board of Longitude'. The report was accepted enthusiastically by the Council, and Wrottesley as President was asked to communicate it to the Prime Minister Lord Palmerston. (He was also to communicate it to the Parliamentary Committee of the British Association, of which he was creator and chairman!) He duly reported that he had done these things, but nothing further appears in the minutes.[33] Sabine nearly twenty years later thought that Wrottesley had discussed the matter with Lord Stanley who had preferred the idea of the Royal Society's becoming the official scientific adviser to the Government. In any case, the matter dropped.

But now in the 1870s the idea was to be revived, although in a different form; it is perhaps a measure of the increased prestige of the Royal Society in government circles, and of the increased importance of science, as well as of its increased professionalisation, that several witnesses, especially Lt. Col. A. Strange (F.R.S. 1864; he had worked on the trigonometrical survey of India and on his return in 1862 become the inspector of scientific instruments being sent from England for use in India), thought it wrong for so much use to be made of the Royal Society's Fellows for what was really government business. Similarly in June 1870 Huxley had told the Philosophical Club 'that he had always objected to the Royal Society giving advice to the Government, and thought it would be better for the latter to request the Royal Society, when its advice was needed, to name committees of experts to consider the question'. Strange and others, including Spottiswoode, preferred the concept of a 'Science Council' whose members would be paid for their work. Strange indeed had a detailed plan which he was delighted to explain at length. It does not seem to be the case that any of those in favour of such a science council or committee thought that the Royal Society had in any way failed in its rôle as adviser to the Government, but rather that they thought it unfair that the Royal Society should do so much work for the Government without any return, and that Fellows should be expected to devote so much of their time to matters of more concern to the Government than to the Society.[34] Thus Strange thought that the proposed council could and should take over the administration of the Government Grant, as well as being responsible for advice to various government departments.

In its review the Commission approved the idea that it was 'a Necessity of the Public Service...to create a Special Ministry dealing with Science and with Education',[35] noting that although such questions had been referred to the Royal Society, expansion would create too much work, and there should be 'a Council composed of men of Science selected by the Council of the Royal Society, together with Representatives of other important Scientific Societies in the United Kingdom, and a certain number of persons nominated by the Government'. (It should be noted that the Royal Society was to have great influence upon this suppositious Council.) In the final recommendations the Commission gave, as number 7, the creation of a Ministry of Science and Education; the next recommendation was for the creation of a council to advise the minister, the council to represent 'the Scientific Knowledge of the Nation' and be capable of dealing with all those matters customarily referred to the Royal Society. Then came a recommendation about the composition of the council – it should represent the 'chief Scientific Bodies in the United Kingdom' and might be given the work of the Government Grant Committee – which strongly suggests that some at least of the objection to the rôle of the Royal Society as the sole government adviser might well have been jealousy on the part of other scientific societies. In the event, of course, there was no new 'Science Council', nor was there to be such a body for many years, and the Government Grant Committee remained a committee of the Royal Society, although enlarged to include representatives of all other leading scientific societies in England, Scotland and Ireland.

The effect, in fact, of the Devonshire Commission's inquiry and recommendation was to strengthen the rôle of the Royal Society as adviser to the Government in scientific matters. For the Commission's report revealed to the general public for the first time the immense utility of a body to advise the Government on scientific matters and the faithful work done by the Fellows of the Society for the Government. This is reflected in the increased number of appeals to it by the Government as the century wore on. The Society responded well. As Lister put it in his 1896 Anniversary address[36]

> We believe that the Council in cordially responding to requests [for advice] and in freely placing at the disposal of H.M. Government its scientific knowledge and its acquaintance with scientific men, is performing one of its most important functions.

As is noted, the Council was 'again and again' asked to approach the Government 'on behalf of the interests of science', and on such occasions 'always meets with a cordial reception and a respectful hearing', even if its request was not acceded to; it was, therefore, he thought, the duty of the Society to respond similarly to government appeals for advice and assistance in scientific matters. This, of course, had been the case over the past fifty years, as earlier. What changed in the course of the nineteenth century was mainly that the good relations between the Society and Government were now obvious, and were seen, as they had not been earlier, as being of mutual benefit.

Relations with other societies

By the end of the nineteenth century virtually every scientific subject possessed its own specialised society. This might, as Banks had feared long ago at the foundation of the Astronomical Society, have meant loss of interest, prestige and power for the Royal. That it did not is the result partly of the inherently inter-disciplinary nature of modern science, partly of the saving common-sense of later officers and Councils, and partly of the fortunate fact that it is extremely difficult for a well-organised and entrenched institution to cease to exist. In the case of the Royal Society no one seriously wished for its disappearance, and the founders of the new societies were almost too conscious of the superiority of the letters F.R.S. to F.X.S., where X is any letter other than R, even when X was ennobled to RX. Only occasionally did the President or other officer have to remind the world that whereas payment of a relatively small subscription and interest secured the Fellowship of other societies, whether Royal Astronomical, Linnean, Antiquarian, Royal Society of Literature or whatever, Fellowship of the Royal Society (entailing a relatively expensive subscription during most of the century) was a privilege, secured only by election by the body of Fellows, and after 1847 only by pre-selection by the Council. In fact, by mid-century most existing societies were only too anxious to be on good terms with the Royal, new societies sought approval of their designation, and gradually all were anxious to cooperate and secure the privileges and power which came to belong to the oldest and, in spite of occasional denigration, the chief among them. By the end of the century, indeed, the paramountcy of the Royal Society had gained it the prestige, power and influence properly belonging to a National Academy, such as most countries, following the lead of France, were coming to possess, without changing in any significant way its private and independent status.

First among peers

No President after Banks seems to have seen any *threat* to the Royal from the formation of other societies, and indeed except for Sussex (and he was a Fellow of the Society of Arts) they were all members of the various specialist societies whose existence Banks deplored, and many had served as presidents of such societies. Although many things conspired to shake the Royal Society in 1830, its essential strength emerges clearly from its triumphant survival of two events in 1831 which Banks would have found traumatic and which he would certainly have regarded as destructive of the Royal's prestige and perhaps of its very existence. These were the transformation of the Astronomical Society into the Royal Astronomical Society and the creation of the British Association for the Advancement of Science. The first of these might appear insignificant, but in fact the granting of the royal charter not only signified recognition of the success of the ten-year-old society, but gave it status in the eyes of the Government. An almost immediate (and very reasonable) result after the accession of William IV was the appointment of representatives of the R.A.S. to be on an equal basis with representatives of the Royal Society to serve as Visitors of the Royal Observatory at Greenwich. With the demise of the Board of Longitude (1828) the way was paved for the Royal Astronomical Society to take over the supervision of the *Nautical Almanac*; at the same time the consultative duties of the Board were now to be shared by the two societies, at the discretion of the government departments concerned. Relations between the societies remained harmonious, especially during the tenure of the astronomical Presidents (Rosse and Wrottesley) at mid-century.

Relations with the British Association might have been acrimonious – indeed have been so represented in the past – for many of those associated with its foundation, especially Babbage and Brewster, were in the vanguard of the attack on the Royal Society during the 'decline of Science' debate of 1830 (see Chapter 2). But neither, in the event, took any leading part in the subsequent history of the British Association during its formative years: Babbage emerged as better at criticism than at constructive action, taking little positive part in the Association's formation and soon withdrawing from participation. Brewster's energies were soon largely absorbed in literary and philanthropic activities; although he attended meetings and supported the organisation in the *Edinburgh Review*, he was not elected a Vice-President until 1840. The anticipated clash between the Royal Society and the British Association – anticipated above all by the Royal Society's critics – never occurred. Instead, there was a merging of interests, amply

indicated by a variety of facts: that every nineteenth-century President of
the British Association was a Fellow of the Royal Society, while a few had
already been P.R.S. and several more were to become so;[1] that most of those
receiving grants for research were F.R.S.; that those actively sponsoring
ideas for the advancement of science in the British Association more often
than not turned to the Royal Society for their implementation; and that the
two institutions were able to work in harmony. The Royal Society's official
attitude was one of benevolent interest. So much was this so that in his
Anniversary address of 1832 the Duke of Sussex could instance the 'noble'
support obtained 'by the eager concurrence of the friends of science' for
the new organisation as a proof that the 'spirit of science' truly existed in
'all quarters of the kingdom', although at the same time he firmly declared
'I believe the scientific character of this country to be most intimately
associated with the scientific character and estimation of the Royal
Society'.[2]

This belief that both institutions were useful and could exist without
friction – provided that the Royal Society's special status were recognised
– undoubtedly reflected the views of the majority of active Fellows,
whatever the minority of the British Association organisers might have
hoped. Reforms in the Royal Society and the natural tendency of institutions
to become more conservative with age brought their views on the
advancement of science into pretty complete harmony, so that, for example,
Wrottesley could happily work for a closer harmony between Government
and science both by creation of a committee of the British Association
(which the Royal Society's Council (17 February 1858) warmly endorsed),
and subsequently by using his prestige as P.R.S., without feeling any
conflict of interest.

Scientific cooperation

It would be tedious to detail all the subjects in whose interest the Royal
Society allied itself with others such as the Royal Astronomical, the British
Association and the Royal Geographical. Some examples must suffice. After
the enforced cooperation between the Royal Society and the Royal
Astronomical Society over the affairs of the Greenwich Observatory in
1830, the next important example that comes to mind – a purely voluntary
one – is the campaign for world-wide magnetic observations. Sabine had
made it quite clear that he turned to the Royal Society to promote magnetic
research, so dear to his heart, because he thought the British Association
lacked any real influence with the Government. Clearly he was wise to do

so, and the Royal Society proved an excellent organiser and leader, backed by the Astronomer Royal (Airy). At the same time the British Association developed its interest while acting in concert so far as possible, although it was not until 1840 that it awarded any money for such research,[3] at a time when the Royal Society's contribution was strong by reason of its influence alone.

In 1843 the British Association was turning to the Royal Society, seeking the Society's influence to secure government money for the publication of Lalande's star catalogue, together with Lacaille's catalogue of southern stars, which had been edited by Professor Thomas Henderson (F.R.S. 1840); as the request was made by Peacock on the recommendation of Baily and of Herschel, it is a little difficult to decide whether this is a case of cooperation between societies, or rather an example of pressure by influential Fellows on the Council of their Society. Indeed in view of the number of Fellows of the Society who were members of the Royal Astronomical Society or of the British Association or both, it seems likely that resolutions were often passed by the British Association, where it was easy to gain purely moral support for *any* scientific proposal (for all might advance science), with the intention by the proposers of exerting pressure on the Council of the Royal Society, which they might well feel would pay more attention to such a resolution than to the proposal of an individual member. Certainly it seems as though such proposals by the British Association as those in 1849/50 for the establishment of a permanent large telescope at the Cape of Good Hope for continued surveillance of the stars of the southern hemisphere (as begun by Herschel) could have come just as well direct from the Royal Society as via the British Association,[4] but no doubt the Royal Society was glad of the extra, 'popular' backing provided by the British Association's approval. Similarly, as noted above (Chapter 5), the inception of the *Catalogue of Scientific Papers* was partly stimulated by publicity given to the idea (itself emanating from the United States) at a meeting of the British Association in 1857. And when in 1866 the Meteorological Committee was organised to superintend the Met. Department of the Board of Trade (see Chapter 5) the Committee appointed consisted of four members of the Royal Society's Kew Committee, and two officers of the British Association (Galton and Spottiswoode), as well as the hydrographer *ex officio* and Col. W. J. Smythe, a noted meteorologist – all F.R.S.

A similar, genuine case of cooperation is to be found in the organisation and running of the Kew Observatory. This had begun as a private observatory in Kew House in the eighteenth century (built by Samuel

Molyneux, F.R.S. 1712) and was used by Bradley; the house subsequently passed into the possession of the Royal Family and the observatory was not usable again. Consequently for the transit of Venus observations in 1769 the King had another observatory (still partly standing) erected in the Old Deer Park, which was maintained by the Government until 1841, when its contents were dispersed. The Royal Society seriously considered taking it over, but although the committee appointed thought favourably of the idea in principle, it also thought that it was too inconvenient in many ways to be desirable, and the Council rejected the proposal.[5] The committee consisted of Herschel, Wheatstone and Sabine, all of whom apparently decided to suggest that the British Association take it over. This was done by raising a subscription, and it was run as a physical observatory by the British Association from 1842 to 1872; at first it concerned itself purely with restricted meteorological research (the humble superintendent was under the supervision of Wheatstone), then with broader meteorological problems, including magnetism, and was available for use by individual members of the British Association, Royal Astronomical Society and Royal Society (including members of its Meteorological Committee).[6] Then, in 1869, the British Association Council decided to sever its connections with Kew three years later, prompting consideration by the Royal Society. After deliberation, Gassiot, a prominent member of the committee set up to consider the matter, offered to establish a fund to support 'physical [i.e. especially magnetic] observations', which were dear to his heart. This generous proposal was accepted, and Kew Observatory continued under the supervision of the Kew Committee, financed by the Gassiot Trust Fund from 1872 to 1899, the only problems arising being in 1894–95 when the director of Kew Gardens (W. T. Thistleton-Dyer, F.R.S. 1880) complained that the name was confusing, and mail, especially from abroad, was apt to be wrongly addressed, causing him much trouble. He wanted the name changed but after due deliberation it was decided not to do so.

By the end of the century Kew was no longer well suited to magnetic observations, being, as already noted, surrounded by railway lines. In the 1890s there was increasing agitation by various physicists for a publicly supported laboratory, especially devoted to quantitative measurement. The British Association Physics Section supported this idea, the Royal Society did not oppose it, and it was determined that such an organisation should be established at Kew Observatory, under the name of The National Physical Laboratory. This was achieved in 1899; the Kew Committee immediately became merely advisory and the Gassiot Fund was now paid directly to the Laboratory, so that the Royal Society's financial responsibility became nominal. Further, the Laboratory soon moved to more convenient and commodious quarters. At the same time the Society retained a strong representation on the governing body (of which the President and officers were *ex officio* members) and virtual control of the executive committee (of which the officers were *ex officio* members while the President and Council were to nominate the ordinary members). In effect, ultimate control was vested in the President and Council, although this was largely delegated, and gradually ceased.[7] It was a happy conclusion to successful cooperation, for the Royal Society had assisted the British Association during its control of Kew with both money and moral support, and in turn The National Physical Laboratory would not have come into being without support from many interested bodies.

No other connection with another institution was quite so closely interwoven as this; as the century wore on there were, however, a number of cases of cooperation, some of which were fairly long-lasting. Cooperation with the Royal Geographical Society was inevitable in connection with various expeditions in the late nineteenth century for which the Royal Society provided programmes, found scientific participants or helped to publish the results (see Chapter 8). In the 1890s the Royal Society so routinely joined with the Royal Astronomical Society in sponsoring eclipse expeditions (so much more important by then because of the development of spectroscopy and of solar photography) that in 1894 a Joint Permanent Eclipse Committee was formed, to last until the mid-twentieth century. Ten years earlier it had been to the Royal Society that the Astronomer Royal turned, hoping to persuade the Council of the desirability of Britain's joining the Metric Convention, or at least taking part in what became the Bureau internationale des Poids et Mesures, and it was the Royal Society's influence with the Treasury and its initiative in arranging suitable terms that secured Britain's adherence, and its continued participation when the Treasury wished to withdraw from it.[8] Also in 1884 the Royal Society had advised the Science and Art Department of the desirability of participation in the International Commission on Solar Studies, again a matter of interest to the astronomers.

Participation in the running of other institutions

The Royal Society had long supplied *ex officio* members on the governing bodies of a wide variety of institutions. Just as it had supplied Visitors of the Greenwich Observatory, so it had supplied trustees of the British Museum from its inception in the mid-eighteenth century. Throughout the nineteenth century the President was expected to serve in this capacity; after 1881, when the natural history section was separated and moved to South Kensington the Society supplied trustees for each institution, as it does today, although now the President is not necessarily expected to serve in this capacity.

In 1888 the Royal Society agreed to provide members of the committee of trustees of the newly formed Lawes Agricultural Trust, acting in concert with the Linnean, Royal Agricultural and Chemical Societies.[9] The Trust was formally set up in 1889 by J. B. Lawes (F.R.S. 1854), with a committee to take over the management of the Rothamsted Experimental Station which he had founded in 1843; this it still does, and the Society still provides three trustees and four members of the committee. Similarly in 1891 the Royal Society joined with various medical institutions in a delegation to the Board of Trade to permit incorporation of the British Institute of Preventive Medicine, a research institution (but also an institution to provide preventive serums and so on against such dread diseases as hydrophobia; it was intended as an analogue to the Pasteur Institute in Paris) which was being threatened by the activities of the anti-vivisectionists.[10] The Royal Society was subsequently represented on its governing body. When in 1903 it became the Lister Institute of Preventive Medicine the Royal Society was named as one of the institutions to nominate one of its seven governors, and it still does so act for this institute which it helped to bring into being. Also in 1891 the Council was informed that the President was to be *ex officio* a member of the governing body of the Imperial Institute; he delegated his power to W. E. Ayrton (F.R.S. 1881), a professor at the Central Technical College, South Kensington, part of the future Imperial College which was in mid-twentieth century to take over the site of the short-lived Imperial Institute, a somewhat sad monument to the imperial glories of nineteenth-century Britain. A different and this time enduring body with whose creation the Royal Society had much to do was the British Academy, officially founded in 1903 in the wake of the Royal Society's refusal to be responsible for representing non-scientific subjects in the international sphere (see p. 196). This was never officially linked with the Royal Society, but was always to enjoy close and cordial relations with it.

Administrative cooperation: housing

As the century wore on, changes in Government and in society created numerous problems only to be solved by close cooperation with other societies and friendly give and take in administrative affairs. These problems were diverse, ranging from relatively trivial matters such as the choice of programme to avoid having topics discussed at meetings on nights when those Fellows most likely to be interested were attending meetings of their own specialised societies, to quite serious problems involving governmental and Parliamentary decisions which affected either the well-being or the conduct of the various scientific societies, where the Royal Society's peculiarly close relations with the Government were of great importance.

This relationship, and the Royal Society's rôle as natural leader in association with the other, specialised scientific societies, was amply displayed in mid-century when the question of housing became acute. The problem arose from a variety of causes. In the wake of the reforms of 1847, some Fellows, notably Grove and Bell, saw the time as ripe for a closer union of the various 'metropolitan' scientific societies, and although true union was thought impossible, 'juxtaposition' was seen as possible and desirable, and the question was much discussed at the Philosophical Club.[11] The first idea was to secure more space in Somerset House (the Royal Academy had moved to join the National Gallery in Trafalgar Square), although London University was known to have sought to secure any vacant rooms; soon Lyon Playfair was suggesting that the then approaching Great Exhibition might produce a useful building. When various societies showed interest the Philosophical Club's committee approached Rosse as P.R.S. and the Council actively considered the matter. It was clear by 1853 that the Royal Society itself needed more room, and that various other societies (the Linnean, Geological, Astronomical and Geographical were consulted) were in favour of juxtaposition, while in the wake of the Great Exhibition there was public feeling that something should be done by the Government for the scientific societies, and, within the societies, a real desire for juxtaposition.

The previous year 'many Fellows' (almost certainly the members of the Philosophical Club) had presented a memorandum to the President and Council expressing strongly the belief that the advancement of science would be 'materially' benefited by the juxtaposition of the Metropolitan Chartered Societies for the Promotion of Special Branches of Natural Knowledge, as they were denominated. The Council authorised the President to approach the Treasury about possible increase in accommodation with a view to 'juxtaposition'; he also approached the then Prime Minister (Lord Aberdeen) who promised to look into the question. Meanwhile formal negotiations with some of the other societies were begun by inviting the presidents of the Linnean, Geological, Astronomical and Chemical Societies to a special Council meeting. (All were of course F.R.S.;

and the officers of the Royal Society were all members of at least one other of the Societies.)[12] All were in favour, although the Astronomical Society hung back, partly because it valued its independence, partly because its officers thought Somerset House too central a position to be given up. But on the whole both the societies and the Government favoured the idea of juxtaposition in 'a Palace of Science' as it was grandly called – until the Government dropped a bombshell by offering land on Kensington Gore, just south of Kensington Gardens (and just north of the future sites of the South Kensington Museums), land left vacant at the conclusion of the Great Exhibition.[13] There had, so Prince Albert informed Murchison, been thoughts of transferring the Royal Academy to that site, but 'as he had been informed, certain London tradesmen objected to transferring it to so distant a position', so that Burlington House was under consideration for its future home.[14] The scientific societies agreed with the London tradesmen that South Kensington was far too remote a 'suburb' (when the Royal School of Mines was ultimately moved there from Jermyn Street Huxley and others felt the same), and decided to reject any such proposal and to urge the suitability of Burlington House, for although London University had secured some accommodation there it utilized these rooms only occasionally, chiefly for examinations.

After much discussion with the Board of Trade, and direct approach to the Prime Minister by, successively, Rosse and Wrottesley, the central portion of Burlington House was finally offered in May 1856 to the Royal, Linnean and Chemical Societies, the Royal Society being charged with the distribution of rooms between the three societies concerned while the Government agreed to provide money to construct a Great Hall to be used for meetings and for the civil service examinations conducted by London University.[15] This proved all very easy to arrange; the Removal Committee negotiated with the other two societies about division of rooms, the cost of furnishing the joint hall, the porter's wages and the cost of tea. The Royal Society moved into its new rooms in the spring of 1857, the first Council meeting being held on 23 April.

The move seemed highly successful: the rooms were adequate and the location convenient. So much was this the case that other societies sought to move there as well (for example the Geological Society, which was told to stay in Somerset House because its rooms were not yet needed), while still other societies began to seek permission (not always granted) to hold their meetings in the Great Hall, the three societies in residence having agreed to consider such questions jointly. But by the summer it appeared that this arrangement might be only temporary, as emerged from a

Parliamentary debate when the Government was subjected to considerable criticism:[16] some M.P.s, favouring the move, objected to its temporary nature as injurious to the well-being of the societies; while at the other extreme some M.P.s objected to the Government's having spent money for the societies' benefit when at least some were wealthy and could and should, it was argued, have provided their own accommodation. Palmerston urged both the desirability of spending Government money on science and the Government's obligation to continue housing those societies which had been offered rooms in Somerset House, some in the previous century, and the affair blew over. The societies settled in, welcoming pressure from other societies to join them. The Royal Society continued to urge on the Government the attractions of juxtaposition, but was careful to assure the Office of Works that there was no room to spare.[17] Its position was the stronger because over the years the three associated societies did allow their rooms to be used by various societies from time to time: for example in 1866 both the Zoological and the Entomological Societies received permission to hold their scientific meetings in the Linnean Society's rooms and the Mathematical Society in the Chemical Society's rooms, all on condition that this be a temporary arrangement; at the same time the War Department had been using the Great Hall for examinations since 1857, and the St George's Volunteer Rifle Corps held their annual meetings there from time to time.

Clearly more room was desirable, but not only did the Government not appear receptive, but in the summer of 1866 as the Treasurer, W. A. Miller, was horrified to learn from *The Times*, it decided to house the Royal Academy in the rooms occupied by the scientific societies, the 'central building' where it still is. Miller's immediate letter of protest to Lord Derby (the new Prime Minister) produced a slightly shame-faced reply from the Office of Works that the Government had the matter under consideration and was about to 'take the advice of Messrs. Banks and Barry', with whom Miller and Sharpey themselves conferred, and from whom it was learned that all the societies housed at either Burlington House or Somerset House were to be accommodated in new buildings on the Burlington House site. A committee was promptly formed consisting of the presidents, vice-presidents, officers and, very sensibly, William Tite (F.R.S. 1835), president of the Royal Institute of British Architects (who would never have been available as a member had he been a younger man); they were charged with drawing up a plan for the new building, which they had completed by the early spring of 1867. But meanwhile the Royal Academy was about to move into Burlington House; what was to become of the scientific societies? The

answer was temporary accommodation in the courtyard, with a little space
in the wing of the old building; there the Society, cramped but active, was
forced to manage as best it could while watching the lawns and trees of old
Burlington House disappear as the new east wing slowly rose.[18] Not until
1873 did the Royal and Chemical societies move into their new and then
ample quarters in the east wing, with the Linnean Society in the south wing,
to be joined in 1874, when the west wing was finished, by its old neighbours
the Antiquarian and Royal Astronomical Societies as well as by the
Geological Society, successful at last in joining with the other societies as
it had long wished to do. Somerset House was now free for government
offices, while Burlington House proved extremely satisfactory (except for
occasional disturbances by the military activities of the Rifle Brigade), even
acquiring electric lighting in 1879.

The close cooperation occasionally envisaged never materialised. There
was never any real sharing of libraries, as had been suggested in the 1850s,
nor did anything come of a plan suggested in the 1890s for formal
cooperation in regard to publication of papers.[19] The most that was achieved
was physical inter-communication between the Royal and Chemical
societies, as the latter had requested when the new building was being
planned (in theory to permit increased social contact on meeting nights) and,
in 1894, a request from the Chemical Society that the Royal Society should
refrain from scheduling papers on chemical subjects on the days appointed
for the meetings of the Chemical Society's council, the request itself
showing that the Chemical Society was run by Fellows of the Royal Society,
which now at last showed no jealousy of its juniors among societies but had
found its rôle as first among peers. So when in 1890 a 'Learned Societies
Registration Bill' was under discussion in Parliament,[20] the Royal Society,
although not itself disturbed, readily agreed to cooperate with other
societies which were alarmed at the idea (the Bill was largely designed 'to
prevent the use by unauthorised persons of any description denoting
Membership or connexion with such Societies'). In the end the Bill was
dropped.

Administrative cooperation: charity

In 1853 and 1855 two Charitable Trust Acts were passed, defining the way
in which charities were administered and putting the surveillance of their
endowments under the Charity Commissioners. Learned societies were not
at first included; then in 1868 it was decided that they were to be counted
as charities, to the alarm of the Royal Geographical Society which appealed

to the President and Council of the Royal Society for guidance and possible action. In this case the senior society seems not to have been alarmed and no action was taken. But in 1879 the Council instructed the President and officers to 'ascertain' whether government assistance might be obtained in order to preclude the possibility of the Society's being brought within the scope of the 1855 Act[21] and hence subject to stringent supervision by the Charity Commissioners and even loss of control over at least a part of its own finances, while in the next year the Society was in fact required to defend its financial activities (involving sale of lands) in the courts in the face of queries by the Commissioners. In 1881 there was a new Charitable Trusts Bill before the House of Lords which, as Evans (Treasurer) informed the Council, contained 'certain provisions which would injuriously affect the Society, should it be determined to be a charity', but, he added, the Marquis of Salisbury 'had kindly consented to bring forward a clause entirely exempting the Society from the operation of the...Act'. (Evidently, as so often, direct access to members of the Government gave the Royal Society a specially favoured position.) The Bill was withdrawn later in the year, no doubt to the benefit of other societies besides the Royal, but it must be said that in this matter the Society seems to have acted only when its own interests were directly threatened, rather than on behalf of all the other chartered scientific societies, as it did on other occasions.

Although the prime purpose of the Royal Society was far from being the distribution of charitable funds, it did on occasion act in a charitable manner. As might be expected, it made small contributions for charitable purposes, as to the Parochial Schools of Mablethorpe in Lincolnshire, where it held land, and to the widows and orphans of its servants. These were private benefactions. But there was (and is) a larger charitable activity, this time involving cooperation with other societies, arising from the foundation of the Scientific Relief Fund at the instigation of J. P. Gassiot in 1859. He prepared a memorandum about 'the establishment of a fund for assisting scientific men or their families when in need of money', which was read to the Philosophical Club, whose members highly approved; it was agreed that recommendations were to be made by the presidents of the Astronomical, Chemical, Geological, Linnean or Royal Societies. Gassiot, who was on the Council, mentioned the matter at the Council meeting of 5 May 1859, modestly attributing the idea to 'several Fellows of the Society', stating that he hoped to raise a permanent fund of at least £2000 to be placed in trust under the Treasurer of the Society. The formal terms, with the names of those prepared to make donations, were accepted at the next meeting (26 May), and so the Scientific Relief Fund came quietly into being, with an endowment rather larger than at first envisioned, which in the course of the century was much increased by legacies (to nearly four times Gassiot's original figure). It was and is administered by a committee, the only change being that in time the presidents of all chartered scientific societies in Great Britain were entitled to recommend recipients. The sums allocated were small, but clearly useful. Some initial tendency to think that the Scientific Relief Fund might be used to pay the annual dues of impecunious Fellows was rendered unnecessary by the establishment of the Fee Reduction (also known as the Publication) Fund which drastically reduced the dues required, and the Scientific Relief Fund was used to relieve cases of genuine hardship.[22] This was a splendid example not only of cooperation between societies, but of the implementation of the old belief that the whole scientific world was a unit of society at large.

International cooperation: towards a National Academy

The Royal Society had, from its inception, looked beyond the national borders of England, corresponding widely with scientists on the Continent and even farther afield, electing to its Fellowship men from all European countries, and often bestowing its medals on men from other lands, while the regularisation of its Foreign Fellows (later Foreign Members) did nothing to lessen the generally held view that eminent contributors to science everywhere should be included in its domain. (There were exceptions to this view: witness Whewell, telling Murchison in 1850 that

> the project of giving all or the principal of our prizes to merit,
> without any regard to the scientific achievements of our own
> Society and our own countrymen, appears to me a cosmopolitan
> claptrap, which, followed out, would make an English Royal
> Society an absurdity.

But he noted that he had 'long ceased to take any part in the business of the Royal Society', and he was clearly out of touch with the opinions of his juniors.)[23]

Inevitably, as the scientific interests of Britain expanded, and as British dominion increased in remote areas of the world, so the possibilities for world-wide scientific cooperation were welcomed and exploited. Foreign societies in both Europe and the United States had long been on the free list for either the *Philosophical Transactions* or the *Proceedings* or both; they now expected to receive such special reports as those published with government money, and most notably the *Catalogue of Scientific Papers*. The immense interest in collecting magnetic and meteorological data not only demanded the establishment of British observatories and observation posts all over the world, but also cooperation with foreign nationals. But all this was occasional and specialised. Broader cooperation with foreign scientific academies and societies only came in the last quarter of the nineteenth century, when these became eager to establish international scientific cooperation. And although the plan for establishing a class of Corresponding Members to be drawn from the whole English-speaking world proved abortive, the existence of such a plan demonstrates clearly the increasingly international outlook of the Society and its ever-increasingly dominant rôle in the scientific world, at home and abroad (cf. Chapter 4).

Mention has already been made (Chapter 6) of the Royal Society's lead in persuading the British Government to adhere to the Metric Convention; the modifications in the Convention as originally conceived by the adhering Continental countries had been achieved by negotiations conducted by the Society with (principally) the Académie des Sciences in Paris, of which many prominent Fellows had been Corresponding Members throughout the century (of the Presidents since 1860, only Spottiswoode was not elected). That was in the 1880s, at the same time that various German scientific academies were endeavouring to turn an Imperial [German] Geodetic Union into an International Geodetic Association, to which it was hoped that the Royal Society might adhere. Nothing came of this at the time, but ten years later, at the moment as it happened when the *Catalogue of Scientific Papers* was becoming an international undertaking (see Chapter 5), attempts were made to broaden the existing association of the academies of Munich and Vienna with the royal societies of Göttingen and Leipzig; these had been meeting annually for consultation for a number of years, and now wished to include non-German institutions.

The Royal Society was the first to be approached, because of its broadly international outlook displayed in its organisation of the *Catalogue of Scientific Papers* and its interest in geodesy and in antarctic exploration (see Chapter 8).[24] (It is of some interest here to recall that forty years earlier, in his Presidential address for 1856, Wrottesley had argued for 'a greater amount of intercourse between the members of the various Scientific Societies of Europe and America', though clearly he envisaged cooperation at a lower level than that practised by the German institutions, where it was an affair of delegates empowered to speak on administrative details.) In November 1896, at the invitation of the German Government, delegates were sent to a Geodetic Union meeting at Göttingen where ideas for future association were discussed; the Council in turn discussed these with interest in the course of the next year, forming a committee consisting of H. E. Armstrong (F.R.S. 1876), Arthur Schuster (F.R.S. 1879, later a Secretary), and the two Secretaries, Foster and Rücker. These considerations were further reinforced by discussions with colleagues at the 1898 meeting at the Royal Society of the now international committee for the *Catalogue of Scientific Papers*. The Royal Society was inclined to join if the organisation were to be 'of a truly international character', and sought and obtained government agreement to the proposal, so that when the committee attended the meeting of the association held at Wiesbaden in 1899, it was possible to discuss detailed plans for the future Association of Academies (the Scientific Cartel, as, following Schuster, it was familiarly called). The Royal Society, by request, took the lead in entering negotiations with the French and other non-German academies, and by November 1899 was prepared to accept the constitution of the new association and to adhere to it, thereby acting as a National Academy.

But now a new difficulty presented itself. The Royal Society was of course a purely scientific body, whereas in other countries the academies and societies combined scientific and non-scientific subjects (in France the adhering body was the Institut, not its constituent, the Académie des Sciences). True to the principles adhered to for the past half-century, the Royal Society refused to consider accepting responsibility for literary and philosophic subjects even by creating a separate section. The Council discussed the problem with certain hastily consulted 'distinguished men of letters', and it became clear that the only solution was the creation of a new institution charged with representing these subjects at an international level; this became the British Academy, intended to be, so far as was possible, an institution parallel in all respects except its history to the Royal Society. The two have retained friendly relations and cooperate on a variety of matters, national as well as international, in all things very remote from the attitude of Banks when the Geological Society was formed.

Education

No discussion of the Royal Society's cooperation with other bodies would be complete without some consideration of its association with educational institutions. In the twentieth century this was of very considerable concern, but in the nineteenth century it must be said the problems of promoting the teaching of and training in science were of minor interest to a Society more concerned with successful scientific achievement than with the formation of scientists or the arousing of public interest in science. Hence the involvement of the Royal Society was more the result of public opinion and Government activity than of internally generated interest.

The middle of the century saw sweeping reforms in education in general, generated by public concern at the stagnation which appeared to exist in the older foundations and a belief that changes in society demanded wide-spread modernisation of traditional educational methods and content. The first such reforms which concerned the Royal Society arose from the setting up of the Royal Commissions of 1850 which resulted in Parliamentary Acts for both Oxford (1852) and Cambridge (1856). Parenthetically one may wonder whether the Society should not have congratulated itself in an age of such drastic attack on established institutions that by conducting its own reforms of 1847 it had anticipated any possibility of enforced reform. A by-product of the Oxford Act was that the Royal Society was in 1856 drawn into the Parliamentary Oxford Commission's plans for the foundation of a new professorship of physiology in conjunction with Merton College (a Linacre Professorship); it was proposed that the President of the Society should be one of the electors.[25] (Ultimately the President was to serve as an elector for a number of Oxford scientific professorships.) Wrottesley, as appears from his Presidential address, was already sympathetic to such involvement, having as P.R.S. continued the pressure for public support of science, by education as well as in other ways, which had emerged from ' the work of his Parliamentary Committee of the British Association, debated in Parliament in the same year. The outcome in Parliament was a motion to appoint a committee 'to consider the question, whether any measures could be adopted by the Government or Parliament that would improve the position of Science or its cultivators in this country', as a result of which the Society's Government Grant Committee was asked to consider the question also.[26] Its report to the Council contained warm support for non-university scientific education, including government aid for the teaching of science. But it can hardly have anticipated that in 1869, the year after the passage of the Public Schools Act, it would be asked to nominate a member of the governing body of most of the great public schools, beginning with two of the oldest, Eton and Westminster. The Society accepted the responsibility, which continues today, without demur.

In the latter part of the century the Society itself showed interest in the problem of scientific education. Partly this arose from the involvement of so many leading Fellows in the activities of the Devonshire Commission, half of whose brief was to consider how to improve such education. Airy in his Presidential address of 1873 referred approvingly to the activities of 'the Official Scientific Commission, of which your Home [*sic*] Secretaries and other Fellows of the Society are members' in reporting on possible methods of improving scientific education as well as scientific research in the universities, and as the Commission's reports show, individual scientists (nearly all F.R.S.) did take the matter very seriously indeed. By the end of the century there was some feeling that the Royal Society should take the lead in promoting the teaching of science generally, even at an elementary level: in 1890 H. E. Roscoe (F.R.S. 1863), who had been knighted for his services to education, brought the question of the teaching of science in elementary schools to the attention of the Council, but failed in his attempt to arouse interest.[27] When in 1897 Foster presented a memorandum to the Council about the teaching of science in the public schools it was agreed to have a discussion between the President, Council and the representatives of the Society on the school governing bodies,[28] but in the immediately succeeding years little was done. Real involvement had to await the Presidency of Sir William Huggins, when possible positive contributions by the Society to elementary and secondary education in science were at last taken in hand and thoroughly canvassed.[29]

8

The encouragement of scientific exploration

The Royal Society had, since its inception, shown a strong desire to assist overseas travellers, in the hope of adding to the body of knowledge possessed by Europeans about distant lands. In the seventeenth century this was mainly shown by the drawing up of numerous lists of queries for travellers to the Near and Far East and to the Americas, the answers to which, it was hoped, would provide extensive material for the universal natural history which so many Fellows saw as the ultimate aim of the Society's endeavour. Natural history meant not simply the investigation of the fauna and flora of distant lands, but included meteorology, hydrography, geology, anthropology, even industrial production. Sea captains could be expected to provide information about sea temperatures, currents, storms and geography, as well as carrying astronomers to observe the skies of the southern hemisphere on the occasion of any special events such as the eighteenth-century transits of Venus (1761 and 1769) – the principal reason for Cook's first voyage which took Banks to the South Seas and established his reputation as a naturalist. In the nineteenth century, first under Banks, then under later Presidents, the Royal Society maintained its interest in voyages of every kind, but particularly in arctic and antarctic exploration, in voyages which might bring in magnetic and meteorological information, in travels in remote parts of Africa and Asia. Towards the end of the century there was a return once again to concern with astronomical ventures.

It must not be forgotten that many Fellows began their rise to scientific eminence by engaging in travel and exploration and the associated scientific investigations. These ranged from army officers like Kater and Sabine to young naturalists like Darwin, J. D. Hooker and Edward Forbes and medical men like Huxley who privately acted as naturalists; while John Barrow, who had himself travelled in the East, was active in the Royal Society during his long career in the Admiralty and was one of the founders of the Royal Geographical Society. The Admiralty connection – which continued after Barrow's retirement in 1845 – made it easy for the Royal Society to be involved in all voyages of exploration or hydrography, while of course the Admiralty had been concerned with astronomical matters since the eighteenth century. The Royal Society often elected as Fellows ship's captains like Ross and Fitzroy who had been involved in voyages of exploration or hydrography for the Admiralty, and in return (as noted in Chapter 6) had often been able to assist its Fellows by securing Admiralty ships for their use in collecting information and specimens. Similarly, through connections with the East India Company, the army and the Colonial Office the Society often had a hand in developing the scientific aspects of travel undertaken for other purposes. By no means all travellers or all important voyages of exploration were connected with the Royal Society, but a surprising number were so connected to a greater or lesser degree, even if only by the bestowal of the Society's Fellowship on the participants.

Arctic exploration[1]

The Royal Society's intimate involvement in the promotion of arctic exploration and with planned scientific studies in high latitudes, like so many of its early nineteenth-century commitments, was a legacy of the eighteenth century. In 1743 Parliament had passed an Act to reward anyone who discovered a North-West Passage from Atlantic to Pacific at a period, it must be recalled, when Greenland had not been circumnavigated nor northern Canada thoroughly explored. Exactly thirty years later and five years before Banks became President the Royal Society first sponsored such a search (under Captain C. J. Phipps, who on his return wrote an account of it) which, sailing to Spitzbergen, was turned back by ice, but did make some observations on magnetism, on currents and on the temperature and depth of the sea. These were all subjects of continuing interest as is apparent from the work of such men as William Scoresby, who was a protégé of Banks.

Scoresby had met Banks after brief service in the navy on its Danish expedition of 1807; he was a captain in the whale-fishery in succession to his father, on board whose ship he had in 1806 reached latitude 81° 31′ North, the highest northern latitude yet attained, and he had also, surprisingly, studied at Edinburgh, as he was briefly to do again. With encouragement from Banks he combined his profession with scientific investigations: hydrography (in 1810 he asked Banks for help in obtaining

Table 1. *The Royal Society's involvement in nineteenth-century voyages of exploration*

PRINCIPAL AREAS OF EXPLORATION

Arctic: 1819–49, 1875, 1882–85
Antarctic: 1838–43, 1845, 1873
Africa: 1840, 1857, 1891

FIELDS OF INTEREST

Geography: 1819–49 (Arctic), 1838–43 (Antarctic), 1840, 1857 (Africa)
Magnetism: 1819–31 (Arctic), 1838–43 (Antarctic), 1845 (Antarctic), 1882–85 (Arctic)
Geodesy: 1821–23 (Sabine, North Atlantic), 1828–30 (Foster, South Atlantic)
Biology: 1838–43 (Hooker, Antarctic), 1872–74 (*Challenger*), 1882–85 (Arctic)
Astronomy: 1874 (transit of Venus), 1875 (eclipse), 1882 (transit of Venus), 1886 (eclipse)

SCIENTIFICALLY SIGNIFICANT VOYAGES NOT SUPPORTED BY THE ROYAL SOCIETY

1821–33	Various ships under both Franklin and John Ross to the Arctic; purpose mainly geographical
1831–36	H.M.S. *Beagle*; South American coast hydrography, with Charles Darwin aboard
1841	H.M.S. *Beacon*; Eastern Mediterranean, with Edward Forbes
1846–50	H.M.S. *Rattlesnake*; around Australia, with T. H. Huxley
1856	Astronomical expedition to Teneriffe, negotiated by its leader, C. P. Smyth, Astronomer Royal of Scotland, approved by Royal Society, but not initiated by it

better thermometers and a 'marine diver' to permit collection of seawater at considerable depth, both for temperature measurement and to collect specimens), botany and magnetism, the last of which continued to interest him particularly; his election as F.R.S. in 1824 came after he had published an account of his voyages and was preparing to enter the Church. Attempting to push ever further northwards on whaling voyages he found that the icepack had retreated, from which he deduced that the climate was becoming milder, and hence that it should be possible once again to sail northwards on voyages of discovery. (He was realistically aware that no North-West Passage could be of any practical utility; his interest appears to have been in establishing geographical facts, with, presumably, a desire to extend the whale-fishery grounds.) He communicated this view to Banks, with whom he was in regular correspondence, and Banks in turn brought the matter up at the Royal Society's Council meeting on 20 November 1817, proposing to write to the Admiralty with the Society's backing to urge the desirability of official voyages of exploration around Greenland. The aims should be to determine whether Baffin Bay, last visited two centuries earlier, really existed, and to search for a possible North-West Passage. Barrow was highly enthusiastic; although, ironically, he doubted the existence of Baffin Bay he was convinced of the existence of a genuine North-West Passage. With such backing the Admiralty quickly espoused the scheme, which so caught the public imagination that a Parliamentary Act was passed to reward with £5000 any ship reaching longitude 100° West (about halfway across the top of Canada), after which the Pacific would seem easy of access. At the next Council meeting (8 January 1818) the President was able to report that not one but two expeditions were to be sent out, comprising four ships, two to sail up each side of Greenland, both to look for western passages to the north of Greenland.

Further, these were to be voyages for scientific as well as geographical purposes. On this point the Admiralty relied on the Society's advice, the matter being in fact dealt with by the Pendulum Committee,[2] which considered the instruments required and provided instructions for their use which were distributed to the ships' officers. But science was to receive more than casual attention, for it was decided that each expedition should have with it 'a proper person' to make 'Observations for the Improvement of Science' in all possible departments. The two men suggested by the Society were Edward Sabine and George Fisher. Sabine, then a Royal Artillery captain, had been living in London for the past two years engaged in scientific pursuits under the joint aegis of Banks and Kater; in anticipation of the results of this voyage he was elected F.R.S. in 1818. Fisher was then an undergraduate at Cambridge but already known to Banks as keenly interested in science; he was to be elected F.R.S. in 1825, receiving his M.A. in the same year, and to combine a career as an astronomer with service as a navy chaplain. Fisher, on the *Dorothea* under Buchanan, was to proceed through the Spitzbergen seas as far north as possible in company with the *Trent* under Franklin; he made a series of important magnetic observations, and observed the effects of iron ships on chronometers. The expedition was scientifically successful, but it was cut short when the *Dorothea* was damaged by ice and forced to turn back.

The *Isabella* (under Ross) and the *Alexander* (under Parry) were more successful: they were to explore Davis Strait towards Baffin Bay, which they duly located, after which they then sailed briefly into Lancaster Sound, which Ross decided (wrongly) was land-locked. On return home Ross was bitterly attacked by Barrow (in the *Quarterly Review*) for not proceeding further, in spite (or because) of the fact that Barrow had doubted the reality of Baffin Bay whose existence was now established. The scientific work of the expedition was excellent: Sabine made a series of important magnetic observations and some observations on natural history, all to be published in the *Philosophical Transactions*, while Ross independently pursued both natural history and oceanography, of especial note being the obtaining of samples of marine life at 1000 fathoms with the aid of his 'deep sea clam'. Unfortunately Ross and Sabine fell out over the question of publication: in his account of the voyage Ross declared that Sabine, whatever his merits as a magnetic observer (which were not in question), had been useless as a naturalist, claiming to know nothing about natural history other than ornithology, and that he had been uncooperative both on the expedition and in preparing the material for publication, charges which Sabine hotly denied and tried to refute.[3] The resulting pamphlet war was vituperative and undignified but it had no lasting consequences, although Barrow's enmity was to restrict Ross's public career for some years.

Once begun, arctic exploration continued vigorously. In 1819 Parry in the *Hecla* commanded another expedition up the western coast of Greenland; although it did not reach as far north as Ross had done the previous year, both Parry and Sabine (on board for more geophysical measurements) returned with enhanced reputations, Parry being elected F.R.S. in 1820. The voyage certainly produced much meteorological and oceanographic information, while Sabine, at the Pendulum Committee's direction, made a series of measurements on a seconds' pendulum designed towards a large-scale plan to determine the exact shape of the earth in all parts of the globe through gravity calculations. Then in 1821 the Pendulum Committee proposed more observations to be carried out by Fisher and Parry in an expedition which twice wintered in the ice and produced more details of the arctic regions. So close was the Society's connection with arctic exploration that when in 1826 Parry put forward a new plan for reaching the North Pole from Spitzbergen it was only accepted by the Admiralty on the specific approval of the Society's Council.[4] On this occasion the scientist was Henry Foster (F.R.S. 1824), a navy captain who had previously served on surveying expeditions off the coasts of North and South America and, in 1823, off Greenland and Norway on Sabine's voyages of gravity determination organised by the Society. Now, with Parry, Foster made an extensive series of measurements on magnetic variation in high latitudes for which he was in 1827 to receive the Copley Medal.

For some years after this the Royal Society showed little interest in exploration, preferring to sponsor voyages of general scientific interest. It was, for example, not involved in Franklin's expeditions of 1819–22 and

1825–27, although it elected him a Fellow in 1825. Nor was it initially
concerned with Ross's privately financed expedition of 1829–33 which
discovered Boothia Peninsula (named after Felix Booth who largely financed
the operation), surveyed large tracts of land nearby (between 90° and
100° West longitude), and during the course of which Ross's nephew, James
Clark Ross, discovered the North Magnetic Pole (1831).[5] But in 1832 the
Council gave a generous grant from the Donation Fund of a year's dividends
to the trustees of the Arctic Land Expedition, which sent out Captain
George Back to rescue Ross if he needed it. He did not need rescuing,
returning safely in 1833, leaving Back to undertake further, mainly overland,
arctic exploration. Ross was not, however, able to obtain any support from
the Society in 1846 when he approached it directly on behalf of his proposed
search for Franklin (missing for a year), a search which he thought of
combining with measuring an arc of meridian off Spitzbergen, since the
Council decided that it would not be 'proper' to give an opinion unless
approached directly by the Admiralty. The case was little different three
years later (and Franklin had then been missing for four years) when in
November of 1849 Lady Franklin wrote personally to Lord Rosse; when
he, as P.R.S., put the matter to the Council, it merely authorised Rosse to
approach the Admiralty to urge its consultation with naval officers familiar
with the problems of arctic exploration (presumably with John Ross in
mind), a very moderate recommendation. In fact the Admiralty officially
decided that no more lives should be risked 'in the investigation of the North
west Passage' and it was left to Ross to mount a private, unfortunately
unsuccessful expedition on behalf of Lady Franklin, Franklin's fate being
finally determined only five years later. There is no doubt that the loss of
Franklin's whole expedition, including the two ships specially strengthened
against ice, the *Erebus* and the *Terror* which had already survived the
Antarctic, deterred any keen interest in further promotion of arctic
exploration for some time, at least within the Royal Society, especially since
Barrow had ceased to be a Secretary of the Admiralty in 1845.

It was in fact to be over twenty years before the Royal Society once again
took an interest in the far north. The impetus then came from the Royal
Geographical Society which, in 1873, approached the Royal Society with
a plan for careful exploration of the area around the North Pole. The
Council readily appointed an Arctic Sub-committee to work jointly with
a similar Geographical Society committee and was willing to authorise a
deputation to the Government to press for official assistance, while leaving
the initiative to the junior society. The result was an expedition sent out
in 1875 with the aim of approaching the North Pole via Greenland, armed

with detailed suggestions and a manual for scientific research drawn up by a Royal Society committee (appointed in December 1874). This expedition achieved little, but it perhaps kept alive interest in the Arctic. For when in 1882 the Russian Government put forward proposals for international cooperation in purely scientific research in the far north the Royal Society was prompt to appoint a Circumpolar Committee, which enthusiastically endorsed the Russian plan to investigate the scientific aspects of 'the great inaccessible region surrounding the pole' by means of groups from various countries who should coordinate their observations of physical and biological phenomena. The British Government acceded to the Society's request, providing £2500 on condition that the Royal Society assume charge of the British team as it had for the past decade done for joint astronomical expeditions. The Circumpolar Committee was in charge of planning and, later (1886), of the printing of the results of the British expedition. This was the last arctic expedition of the century to involve the Royal Society, but in 1899 it was as ready as ever to join the Royal Geographical in forming a Joint Arctic Committee set up to plan a new, twentieth-century attack on the Pole.

Antarctic exploration[6]

Until the late 1830s the Royal Society's interest in the southern hemisphere of the earth was chiefly confined to the advocacy of the establishment of fixed astronomical and magnetic observatories, first in South Africa at the Cape and then in Australia (especially at Melbourne), and it displayed little interest in the possibility of observations made in high southern latitudes to complement those made in northern latitudes. When Foster in *Chanticleer* reached the South Shetland Islands (see p. 214) there was no demand for pushing further south into the Antarctic Circle, nor had Weddell's accounts of his whaling voyages into that circle earlier in the 1820s aroused anything like the response which Banks had given to Scoresby's northern voyages. It has been suggested that this lack of interest reflected Barrow's, for he appears to have been exclusively concerned with arctic exploration. By way of contrast, Beaufort, an active navy captain who became hydrographer to the navy in 1829, was particularly interested in southern exploration, and it was he, rather than the Royal Society (of which he had been a Fellow since 1814, but to which no reference was made on this occasion), who was responsible for promoting the scientific side of Fitzroy's *Beagle* expedition, being indirectly responsible for Darwin's presence.[7] Beaufort was always anxious to promote the scientific side of any official expedition; he was also, not surprisingly, one of the reformers in the Royal Society itself, siding with the rebels and those who wanted the Society to be more exclusively scientific, although he was much older than most of them (see Chapter 3).

Beaufort, indeed, must have had much to do with the Royal Society's attitude in 1838 towards the proposal for J. C. Ross's voyage to high southern latitudes, the first great official voyage of exploration in antarctic waters. Ross had been a Fellow since 1828, and had been active in magnetic affairs; he had discovered the North Magnetic Pole in 1831 and in 1838 was employed by the Admiralty in a magnetic survey of the United Kingdom. Not surprisingly, he was highly thought of by the Society's Council, and especially by Sabine, who had been with him in the arctic expedition of 1818 under John Ross, and by John Herschel, recently returned from his astronomical surveys at the Cape and now honoured by the Government with a baronetcy. Both saw the possibilities inherent in a voyage to high southern latitudes for the grand study of magnetism so dear to, especially, Sabine, who was busily promoting the cause of magnetism within the British Association, and both saw J. C. Ross as the ideal man to lead such an expedition. As the Presidential address of 1838 put it, 'We are rapidly approaching great and comprehensive generalizations, which can only be completely established or disproved by very widely distributed and, in many cases, by absolutely simultaneous observations', and this was naturally particularly true of magnetism and meteorology.

Fortunately, the Hydrographic Service was keenly interested in all scientifically orientated surveys.[8] The only real problems were the closely connected ones of money, time and influence. It was necessary, if an expedition was to be successfully mounted, that the Admiralty be prepared to ask Parliament for the requisite funds, and that it be prepared to release a suitable officer to command the ship or ships. For the first requisite the times were decidedly unsuitable, for Melbourne's Government was unstable during the spring of 1839, although not yet in the autumn of 1838 when plans were first being made; more important in terms of time must have seemed the fact that Ross was just completing his tour of duty in the magnetic survey, was eager to lead the proposed expedition, hoping to reach the South Magnetic Pole as he had the northern one, and if the expedition were not quickly decided upon he would have to accept another command. Timing was also critical in that suitable ships were required; further, the high southern latitudes could only be accessible in the southern hemisphere's summer, and several months were necessary for the journey. Sabine (whose chief fear was that Ross would not be available) planned the scientific details with Ross and Beaufort as the Royal Society's acknowledged leader in magnetic affairs, urging Herschel to apply his influence on the Government while Whewell was asking the Royal Society's Council 'to strengthen the application from the Association'. Timing was to be more critical than Sabine could know, for Melbourne's Government fell only a month after Ross received his appointment (in early April 1839), and although Peel, sympathetic to science generally, was expected to succeed him, what mattered was the interest of the First Lord of the Admiralty; in any case there would almost certainly have been delays during the change of Government.[9]

Although planning had begun in the autumn of 1838, it was not until the very end of September 1839 that the *Erebus* (Ross's ship) and the *Terror* (commanded by another experienced explorer, Crozier), both specially strengthened for passage through ice, set sail on their three-year voyage. Their instructions were detailed and explicit: Beaufort had seen to it that the Admiralty instructions stressed scientific observations and the Society's Committee of Physics and Meteorology (of which Herschel was chairman) drew up a hundred-page volume of magnetic and meteorological procedure, while there were lesser 'catalogues of desiderata' drawn up by the Geological, Zoological and Botanical Committees. The expedition was the best equipped and prepared yet mounted, far more concerned with science than any earlier such exploring expedition. True, its principal scientific aim was the study of magnetism which 'must' (as Ross noted in his *Voyage of Discovery*) have eventual practical applications to navigation, but the motive of the voyage was the search for knowledge. Ross acted not only as captain but as principal scientific officer, overseeing both magnetic and meteorological observations with great care.

The resultant scientific investigations were of undoubted value. Of particular note is the fact that Ross was able to arrive off Kerguelen Island early in May 1840 so as to fulfil the plan made beforehand for careful magnetic observations to be made there on 29 and 30 May for comparison with those to be made on the same day at all British and foreign magnetic observatories. As it happened, those were days of unusual magnetic activity at Kerguelen Island so that Sabine and the Committee were particularly pleased. This success was in some measure to offset the disappointment, keenly felt by Ross, at his inability because of impenetrable ice to reach the South Magnetic Pole, calculated by Gauss as lying at 72° 35′ South, 150° 30′ East; Ross calculated that the ship had reached to within 150 miles of it, however. Kerguelen Island was a double success, for it also supplied important botanical observations made by the *Terror*'s assistant surgeon, the young and untried Joseph Hooker (see below).

Geographically the expedition broke totally new ground. After sailing to Hobart the ships in 1840 set out on a five-month cruise mainly within the Antarctic Circle. They met the hitherto unknown Great Ice Barrier (later named after Ross), having in January 1841 reached the Ross Sea and discovered Queen Victoria Land of which Ross took formal possession. The expedition roughly surveyed the land from latitude 72° to 79° South, far farther south than had been reached before, and establishing the existence of land much farther south than had been known to exist. They then returned to Hobart, Ross wisely deciding not to risk wintering in the ice. In November 1841 he once again sailed along the Barrier westward across the Ross Sea, probably sighting King Edward VII Land, as it was subsequently called, and continuing on to the Falkland Islands. On the third voyage, 1842–43, Ross charted part of the eastern coast of Graham Land in the course of penetrating deep into the Weddell Sea, reaching latitude 71° 31′ South and longitude 14° 51′ West, returning home via the South Shetlands and St Helena. Ross had sent home news of his discovery of the Great Ice Barrier and of having reached latitude 78° South; much to Beaufort's annoyance the Admiralty refused to publish a notice of these discoveries, which had to wait until Ross's return when he himself published an account (1847), while his magnetic observations were published by Sabine in the *Philosophical Transactions* (1842 ff). Northampton's Presidential address for 1842 (printed in *Procs.* IV) contained a useful report on the expedition as a whole, from which, as the address for 1843 recorded, Ross was to bring back 'valuable scientific details'. Altogether it was exactly the kind of expedition deserving of the support which it had received from the Royal Society.

So successful had Ross's voyage been that there was little official interest in further antarctic exploration for several decades. There were expeditions for astronomical purposes, notably that to observe the transit of Venus of 1874, which went to moderately high southern latitudes (see below), but there was little concern for further expeditions to the Antarctic Circle until as late as 1887, when the administration in Victoria (Australia) approached the Colonial Office about the possibility of money for a new expedition to Antarctica to promote trade and for scientific enquiry; the Colonial Office in turn asked for advice from the Royal Society. Both the Colonial Office and the Royal Society's Antarctic Committee doubted whether the initiators understood the scale of support that would be necessary, and although the Royal Society was in principle in favour of the proposal, nothing further came of it. As with arctic exploration, there was renewed interest and concern only at the very end of the century, initiated by the Royal

Geographical Society but warmly taken up by the Royal Society as well. Within the Royal Society the moving spirit was P. L. Sclater (F.R.S. 1861), a zoologist noted for his work on the birds of Argentina, who, three years after the original interest shown by the geographers, proposed and secured a discussion meeting on the subject held on 24 February 1898. The scientific basis for the great expeditions of the early twentieth century was thus firmly laid, well before the next stage of antarctic exploration associated with such names as Scott, Amundsen and Shackleton.

Exploration by land

There is no doubt that it was the great sea expeditions which not only seized the imagination of the public but which were of great potential interest to science. But there were very many parts of the globe where the land was little known to nineteenth-century Europeans, and this in all the continents, even including a few parts of Europe. The largest unknown area (aside from Australia, being explored and colonised simultaneously) was the interior of Africa, which also posed a very great number of geographical puzzles, above all the origin of the great rivers which flowed to the coast. The Royal Society was briefly associated with a number of African explorations. Thus in 1840 when Captain H. D. Trotter was preparing to take a naval expedition to the Niger, to explore and to make treaties with the local chiefs, he was told by Lord John Russell that he must consult with the Royal Society about the requisite magnetic instruments, a forcible indication of the widespread appreciation in official circles of the magnitude and importance of the global investigation of magnetism. (Trotter was given prompt help by the Physics and Meteorology Sectional Committee.) Unfortunately, as Northampton had to report to the Society in 1842, this expedition was not very successful in terms of scientific results, except ornithologically speaking. Then in December 1857 the Foreign Office sought the Royal Society's advice about Livingstone's proposed explorations in central and eastern Africa (he was about to be elected F.R.S. for his earlier discoveries); a committee conferred with him and ascertained that he was to be accompanied by a botanist and a geologist, and supplied him with advice from Hooker, Owen and Murchison as to the kind of information most desired. Nothing else equalled the interest in Livingstone's explorations, although in 1888 the Council was willing to respond to a request for support for a proposed expedition to the Atlas Mountains.

By this time the Society's interest lay rather in the medical than the magnetic or geographical aspect of African exploration, and in any case proposed expeditions, like that to the Karakoram Mountains in 1891, usually included men who knew about the Society's resources and applied for a Government Grant to cover any scientific expenses when there were real scientific objects in view.

Expeditions of biological importance

While magnetic inquiries tended to dominate the Royal Society's concern with all expeditions, by land and sea, at least for the first half of the century, it was by no means the case that biological considerations were overlooked. Although not so much stressed they were taken seriously, and by the last quarter of the century were very seriously taken indeed. Even on John Ross's arctic voyages, when the prime consideration was geographical exploration and the second consideration was magnetic observations, some attention was paid to biological observations. There were of course few plants to observe, but Ross's *Narrative* had included accounts of zoology (by J. C. Ross), geology (by John Ross), and much on the Eskimos' languages and culture. (As noted above, Sabine was officially the expedition's naturalist, but as Ross scornfully remarked he was not much interested in collecting, professed no competence in anything except ornithology, and preferred to publish his accounts separately in any case.) Clearly the biological possibilities of the voyage were not much emphasised either in planning or in execution; except for a little oceanography (current flow, sea temperature and depth) in fact little but magnetism was of concern. Only later was oceanography combined with biological interest in that measurement of the depths of the sea was combined with the collection of marine life at varying depths.

The first expedition in which the Royal Society showed any interest in systematic collection of plants and animals – for, as already observed, the *Beagle* voyage seems only to have come to the Society's attention after its return – was Ross's antarctic voyage. The naturalist was Robert McCormick, a naval surgeon who was later to be well known for his accounts of various voyages (he was also to accompany the 1852 search for Franklin), whose interests seem to have been zoological; he was on the *Terror*. Aboard the *Erebus* was the young Joseph Hooker (just twenty-one and barely qualified as a surgeon), given the title of botanist by Ross; in fact although his primary interest was botany he acted as general naturalist and marine zoologist, being without competition on his ship, and often allowed ashore. The Committee of Botany and Vegetable Physiology, of which John Lindley (F.R.S. 1828), professor of botany in the University of London, was chairman, was asked by the Council to draw up a 'report' (really a list of recommendations and advice) for the expedition, formally presented to Ross and Hooker in early July 1839 when Thomas Horsfield (F.R.S. 1828), keeper of the East India Company's Museum, Jonathan Pereira (F.R.S. 1838), pharmacologist, and J. F. Royle (F.R.S. 1837), surgeon and botanist, professor of materia medica at King's College London, represented the Committee.[10] Hooker, with the arrogance of youth, thought that there was 'little new in... the long list of advices' which they gave him, and could not even get Horsfield's name correctly; he thought them not 'cordial... in the least degree', and in the course of his work seems to have ignored the instructions completely. As was true of Darwin, a long voyage was the school Hooker needed: at sea he learned marine zoology; ashore, in Tasmania, New Zealand, the Falklands, South Shetlands and Kerguelen Island (where he paid particular attention to the local variety of cabbage, first discovered by Cook but never before properly described), Hooker botanised extensively and the results were to appear in Hooker's own *Flora Antarctica* (1844–47).

While the Royal Society's view had not much influenced Hooker, he had influenced it, for he was to be elected a Fellow in 1847, and this voyage encouraged the Society's later support for a biological side to exploration. That Hooker was not alone in surmising a certain lack of interest within the Society at mid-century in the biological aspects of exploration is suggested by the fact that Edward Forbes was recruited for his voyage to the Mediterranean in 1841 by the independent action of the ship's captain who invited him to accompany the *Beacon* on a surveying and dredging voyage; although he was shortly afterwards (1845) to be elected F.R.S. it does not seem that the Royal Society was aware of his travels at the time they occurred.[11] But thirty years later this attitude had changed. With Hooker then President, a great deal of attention was paid to the necessity of having naturalists attached to the transit of Venus expeditions in 1874: there was a special committee, by whom two men were chosen, one for Kerguelen Island, and one for Rodriguez Island, who were paid to make collections and given detailed instructions, and whose reports were carefully reviewed by the committee before publication.

But the most important biological expedition, whose main purposes were marine zoology and oceanography, was the voyage of the *Challenger* 1872–74.[12] As already noted (Chapter 6) oceanography and its associated study of life at various depths of the ocean was a particular interest of W. B. Carpenter and of Wyville Thomson. Besides Thomson, the *Challenger* under Captain G. S. Nares (F.R.S. 1875) carried two biologists:

John Murray (F.R.S. 1896), ultimately chiefly responsible for the monu-
mental *Report* (1880–95), and H. N. Moseley (F.R.S. 1877). There was also
an oceanographer, the navigating officer T. H. Tizard (F.R.S. 1891 when
assistant hydrographer of the navy) and a chemist and geologist, J. Y.
Buchanan (F.R.S. 1887). There was no other expedition like it, none so
thorough or so extensive, none so closely associated with the Royal Society
(partly because of the support given by the hard-working biological
Secretary (1872–81), Huxley).

After crossing and recrossing the Atlantic at various latitudes the ship
went from Cape Town south to Kerguelen Island and then southwards into
the Antarctic Circle, the first steam-powered ship to do so, and thence to
New Zealand and across the Pacific, with particular attention to the western
part of that ocean before rounding South America via the Straits of
Magellan. The results of the voyage were to appear slowly for the remainder
of the century, carefully supervised by the Royal Society's Challenger
Committee, so long lived as to be almost a standing committee (formed in
1871 as the Circumnavigation Committee it was reappointed for the last
time in 1898). The programme's chief activity was dredging in the thorough
manner initiated by Carpenter and Thomson: sounding, taking water
temperatures and samples for analysis, dredging the bottom, and retrieving
marine life at great depth (up to 3000 fathoms). The scientists found much
information about the composition of the ocean floor, about migration of
plankton, about the distribution of species, and about temperature variation,
but lacking a physicist they were able to do little work on the nature of ocean
currents and the circulation of the waters. But it was altogether a tremendous
undertaking, which was to provide raw material for further, early twentieth-
century, investigations. There was no other expedition like it for scientific
importance, except possibly Ross's antarctic voyages, in which the Royal
Society's rôle was of almost equal importance.

Astronomical, geodetic and magnetic expeditions

No expedition, even that of the *Challenger*, was devoted to a single science,
for the opportunities for varied forms of research are too obvious to need
emphasis. Indeed, it will be evident from the discussion above that
specialisation was not regarded as necessary or even desirable until the last
part of the century, most of the scientists carried being called 'naturalists'
at best, and at worst, like the young Huxley on the four-year surveying
voyage of the *Rattlesnake* in Australian waters, being surgeons or ship's
officers whose scientific observations were their own affair (the official

naturalist was John MacGillivray who had already served in this capacity on previous voyages). Nor was the Royal Society concerned in *all* the many expeditions and surveys of the century, even when the ships carried scientists aboard. Generally speaking the Society was only involved with major undertakings, where its advice was needed to decide on the scientific advantages to be gained, or where the motivation of the expedition was primarily scientific, like the voyage of the *Challenger*. Except for that and certain geographically orientated voyages early in the century, the prime motives were astronomical, geodetic or magnetic investigations.

It was natural for the Council to show interest in the expedition to Teneriffe made by C. P. Smyth, Astronomer Royal of Scotland, with the intention of utilising the Peak to get above the lower layers of the atmosphere and so gain better observations (perhaps the first suggestion which was to lead to the modern concept of observatory siting). A few suggestions were made for subsidiary observations when the Admiralty, which paid for the expedition, approached the Society for comment and Smyth was elected F.R.S. in 1857, on his return. The great voyage on which Banks had accompanied Captain Cook had been conceived as a method of getting astronomical observers to a point on the earth's surface where the expected transit of Venus of 1769 would be fully visible; this was not only an event of great astronomical interest in itself but one which could be used to determine the dimensions of the solar system, notably the distance of the sun from the earth, with greater accuracy than had yet been possible. Any geographical and biological discoveries were regarded as subsidiary to the main purpose of the expedition, although as things turned out they were of more lasting importance than the purely astronomical observations. There were no further transits of Venus to observe for over a century, but when these were at last due (1874 and 1882) the Royal Society was still interested. The principal astronomical activity was the concern of the Royal Astronomical Society, but when in October 1873 the Council discussed plans for the expedition, in the mounting of which it had agreed to work with the junior society, it was decided to recommend the inclusion of naturalists and geologists. Hence arose, as already noted, the Royal Society's responsibility for the naturalists on Kerguelen and Rodriguez Islands in 1874. It also took some part in securing money from the Treasury for both the 1874 and 1882 expeditions, through a Transits of Venus Committee, which assisted in organisation.

Soon after this, solar physics, spectroscopy and solar photography became of importance, and hence every eclipse of the sun became the subject of intense scientific interest with a concomitant demand for eclipse expeditions. These were of course always joint expeditions with the Royal Astronomical Society, and while the appropriate committees for the 1875 and 1886 eclipse expeditions were *ad hoc* committees of the Royal Society, in 1894 a Joint Permanent Eclipse Committee was set up, for the eclipse in 1896 and all subsequent eclipses: this included a representative of the Hydrographic Department of the Admiralty, which much assisted preparations.

Geodetic affairs were mostly of concern early in the century. Besides the trigonometric surveys of Great Britain (including Ireland) and India, in all of which the Royal Society participated to some extent, the Society was more deeply involved in attempts to establish the exact shape of the earth by means of careful measurements of the length of a standard pendulum exactly beating seconds at various places on the globe, a method by which the general shape of the earth had been determined at the beginning of the eighteenth century. Sabine had been chosen by the Society to engage in such an expedition on his return from Parry's arctic expedition of 1819–20; the initial voyage (1821–23) was to various stations in the South Atlantic and the second in 1823 to several stations in the North Atlantic, from New York to Spitzbergen. Five years later Henry Foster was given command of his own ship, the *Chanticleer*, and under directions provided by the Pendulum Committee was to sail extensively in the South Atlantic and Pacific Oceans on the same mission, a work he carried out with great care. It was in fact on the conclusion of this work, in 1830, that Foster, after determining the difference in longitude between the two ends of the Isthmus of Panama, was accidentally drowned in the Chagres River. His magnetic observations were considered of great importance to science, and as the Council report for 1831 noted, they were entrusted to Christie (then a Secretary) to condense for publication.

Most magnetic work was, as indicated above, done in connection with geographical exploration or, in the case of Foster's *Chanticleer* voyage, in conjunction with geodetic work. In one single case an expedition was mounted for purely magnetic purposes: this was the voyage of the *Pagoda* in 1845, mounted by the Admiralty at the request of the Royal Society, a request instigated by Sabine through Herschel and the Committee of Physics.[13] On board the *Pagoda* was Lt. Henry Clark, R.A., who had been in charge of the magnetic observatory at the Cape of Good Hope; the *Pagoda*, hired by the Admiralty but under naval command, was, as reported in the Presidential address for 1845, 'to complete the magnetic survey of the high latitudes of the southern hemisphere, of which three quarters had previously been accomplished under Sir James Clark Ross'. To this end the *Pagoda* sailed from the Cape in January south along the Greenwich meridian until it was stopped by ice in the vicinity of the Antarctic Circle, and then coasted along the ice eastwards to latitude 120° East, making magnetic measurements all the way, thus reaching the point to which Ross had sailed and rendering the magnetic survey of the region of the Antarctic Circle complete. Simultaneously there were surveys being conducted from India and Singapore, so that by the 1850s the initial survey originally planned by Sabine and his friends and sponsored jointly by the Royal Society and the British Association was virtually complete. Thereafter, although magnetic observatories continued to function at various points on the globe, supervised by the Kew Observatory under the aegis of the British Association, there were no more exciting voyages, and indeed after the middle of the century the magnetic fervour slowly declined. However, when at the very end of the century arctic exploration revived it was in connection with magnetic research, although now of an international kind.

Clearly, the Royal Society's encouragement of exploration during the nineteenth century had produced a great deal of data for scientific study in a whole range of fields, as well as increasing geographical knowledge. The great expeditions of the early twentieth century in which it played some part, while still retaining some element of pure competition to survey the unknown, with an element of personal and national rivalry to be the first to reach the Poles, had profound scientific bases, demanding a wealth of equipment, a large team of experts, and a deeply sophisticated approach to the possible fruits of exploration to high latitudes or to remote regions of the earth. But the spirit differed little from that displayed from the early part of the nineteenth century to its end.

The end of the century: a truly scientific society

Any institution which is successful takes on a life of its own, and the Royal Society is no exception. Its corporate presence is in itself a phenomenon of interest not only to its component individuals – officers, Fellows, staff – but to the literate public. And in the nineteenth century, though hardly earlier, this was even more the case than it is today. Hence the public excitement when things went astray. But its mere existence as a self-perpetuating society to which it is an honour to belong is not its main *raison d'être*; were it so its history would not be of interest except to its members, and it would never have survived to celebrate its tercentenary in the mid-twentieth century in full vigour. What was of importance was, and is, the purpose for which it was founded, a purpose embodied in its title, 'for improving naturall Knowledge'.

Until the nineteenth century there was no doubt about what 'naturall Knowledge' was, nor any feeling that the Royal Society was failing in its purpose. But during the first third of the century the emergence of the word 'science' – in exactly the sense in which it is employed today in English, but in no other European language – coincided with the need for reappraisal of the Royal Society's rôle in the world of learning, and with a power struggle within the Society. After Banks's death the Royal Society might, conceivably, have gone the way of Continental Academies, its Fellows supported by the Government and its concern with the world of nature limited to one section or division of the whole. That it did not do so is a remarkable phenomenon, all the more remarkable because it retained its private nature, its peculiar relations with Government and its prestige, while expanding its activities to meet the need for a National Academy.

The limitation of numbers which took place after 1847 would not by itself have effected the change from a Fellowship composed of a mixture of serious scholars and dilettanti gentlemen, with many of those scholars not scientists in the English sense but archaeologists, numismatists and antiquaries, to a Fellowship limited in principle to contributors to science as judged by their achievements, which were required to be original and important in the eyes of their elders, already Fellows. The Fellowship was to be composed of an elite, and this elite was to be limited to scientists whose work was of a sufficiently broad scope to make them eminent beyond a narrow discipline. The aim of the 1847 reformers was to create a truly scientific society, and their decision to limit numbers was one means to that end, but not the only one. The Royal Society owes more than is normally realised to Wrottesley, an amateur in the modern sense, although a very serious one, who used his presence in the House of Lords to good advantage. It was for the British Association, not the Royal Society, that he formed and led a Parliamentary Committee – opinion in the Royal Society would have felt that there was no need for an intermediary between the Society and Parliament – from which sprang the remarks in his 1856 Presidential address about the Committee's inquiry into methods of improving 'the position of Science or its cultivators'. This in turn produced the first suggestion for a Board of Science more seriously discussed some fifteen years later by the Devonshire Commission. Wrottesley thus played an important part in educating Parliament; his ideas were also valuable both in helping to convert the less purely scientifically orientated Fellows of the Society and in persuading the 1847 reformers that the leadership of the Society was favourably inclined to their cause.

It is interesting to compare the attitudes of such would-be reformers as Babbage in 1830 with such genuine reformers as Grove forty years on: whereas Babbage and Herschel saw a Government-financed Academy on the model of France as the ideal, in 1870 (during a discussion at the Philosophical Club which, it should be remembered, was especially founded in 1847 to preserve the reforming ideals of that year) Grove declared that he had a 'strong objection' to government support 'for anything in the shape of Academies', instancing the Continental experience which, he believed, 'tended to foster antiquated notions, and to repress the free expression of opinions'. This is a view to which most of the scientific supporters inclined. And, more dramatically, in 1888 Grove, a leader in 1847 (now admittedly an old man), could remark that the Philosophical Club was no longer needed as an entity distinct from the Royal Society Club, because *all* the Fellows were then scientific. Although the merger was not to be effected finally until 1901 this delay appears to have been the result of custom rather than disagreement with Grove's view, for no one could doubt that even before 1888 the Fellows *were* all scientific.

Even more important than the fact that this was so is that by this time there was no disagreement among the Fellows that it ought to be so, that there was now no one who wished to reverse the trend towards a purely scientific society. There might be – there almost certainly was – feeling from time to time that some successful candidates were not as eminent as they might be, or that some who did not succeed were as worthy as those who did succeed. The strict limitation in membership, so that only fifteen candidates were elected in each year, made it a common occurrence for even ultimately successful candidates not to be elected at their first being put up, as the number of candidates in each year increased dramatically, doubling between 1860 and 1900.[1] Not until 1930 was there a modest increase in the number of Fellows elected in each year, and even in 1978 it was only forty, making a Fellowship of 853 (759 ordinary Fellows), as against 613 (531 ordinary Fellows) in 1800 and 496 (449 ordinary Fellows) in 1900, not a great increase considering the vastly increased scope of science and the very large number of practising scientists, as well as the enormous increase in the population.

All this being so, one is justified in viewing the history of the Society during the nineteenth century as being in large part the history of the triumph of those who struggled to create a Royal Society all of whose Fellows should be eminent for their contributions to science. Eminence in other fields and interest in science was now not enough, nor was professional devotion to science, without eminence. The final, irrevocable decision to retain this commitment was made in the late 1890s, when adhesion to the international organisation then being created necessitated the creation of the British Academy to represent those non-scientific disciplines which Continental language and ways of thought regarded as being in method, although not in content, akin to the natural sciences. But there was no difficulty about the Royal Society's acting as a partial National Academy.

Not that the whole effort of the Society during the nineteenth century was devoted to this purpose, which was rather a means to an end than an end in itself, rather an internal matter than one of proper concern to the outside world, although the outside world of clubs and journals and newspapers was keenly interested. Even as regards internal affairs there were other matters of concern, many of which were of continuing interest and of continuing anxiety well into the next century. Efficient organisation of the offices and library, for example, was and is of importance. Housing – room for books, for officers, for Fellows on their visits, for conversaziones and, of course, for meetings – was and is of concern, and standards were progressively improved and comfort increased.

More important was the question of publication – of the need to publish more of the proceedings of the Society and to publish them with reasonable speed. Hence the decision in 1832 to publish as *Proceedings of the Royal Society* abstracts of all papers read, together with any important minutes of meetings; by 1854 some papers were there published in full, leading to very much larger volumes. Even with this method, which permitted greater selectivity towards the *Philosophical Transactions*, that publication increased so much in bulk that from vol. 178 (1887) the *Transactions* has appeared in two series: A for mathematical and physical subjects, B for biological. Further the distinction between the two publications (*Phil. Trans.* and *Procs.*) was minimised in 1887 when in the list of Fellows those who had published in *Phil. Trans.* were no longer distinguished, a change which also marks the increasing disappearance of Fellows who had not made written contributions to science. The only remaining problem was that of reducing the interval between the receipt of a paper and its publication, still of importance today.

A notable change in the Society during the nineteenth century is its attitude towards public knowledge of its activities. In 1800 it published the *Philosophical Transactions* and that was that. Any other knowledge the public might have came through indiscretion or through attendance as 'strangers' at meetings, introduced by a Fellow. From surviving accounts it is clear that at least some of those introduced had little or no idea of the real character and intention of the Society,[2] although others, of course, did, and held themselves privileged to witness the meetings directly. Indiscretion led to many newspaper reports exacerbating existing dissensions. Slowly the Society learned the value of giving information directly to the 'public press': first lists of Fellows and officers, then abstracts of papers (after 1847 prepared by the author and submitted with his paper), although not permission for journalists to attend meetings unless introduced by a Fellow (no stranger was allowed to attend without an accompanying Fellow until the mid-twentieth century). Then in 1896 the Council decided to publish a *Yearbook*, intended for the information of the Fellows but also as a formal publication available to the public. This, originally a slim volume but now a stout one, contained a calendar of the Society's fixed meetings, a list of Fellows, committees with their membership, statutes, standing orders and regulations for medals, financial statements, and Council reports; and in the first few years it contained the Presidential address as well.

The message conveyed by the report of Council and by the Presidential address had, by the end of the nineteenth century, become almost standardised, because the aims of the Society were clear to all. In 1896 Lister (representing the B side) stressed the responsibilities of the Society as an adviser to the Government; in 1980 Todd (representing the A side) did the same: he declared, and the Council endorsed his view, that the Society's first object was and must remain to protect and encourage science in all its aspects, pure and applied.[3] This is an aim with which all nineteenth-century Presidents would have agreed. But only in the latter part of the century would they have given to the words exactly the same meaning implied in 1981. In so far as the Fellows of the Royal Society regard it today as a truly scientific society, it may be said that it became so because of the constant desire of many of its ablest Fellows during the nineteenth century that it should become so.

But it is still worth recalling the struggle to achieve this, and to note that scientific eminence did not necessarily confer automatically on all the prescience required for the necessary reforms. The Royal Society owes a decided debt to a number of Fellows and officers who cared deeply for the advance of science and helped to make the Royal Society a worthy place for its practitioners, without themselves necessarily having made important contributions to direct scientific advance. Nineteenth-century society was vastly different from late twentieth-century society, and vastly different talents were needed in those required to preside over a national society and to approach politicians and government officials. This has been an account of some aspects of the past, which has attempted to understand men's motives and actions in historical terms. Similar though the apparent aims of the actors in this history were to modern aims, their motives and actions were necessarily not modern ones; viewed through modern eyes it is difficult to be sure of understanding them. It is to be hoped that these nineteenth-century Fellows seem interesting if not in all ways admirable predecessors of the Fellows of the Royal Society of the present day.

A NOTE ON SOURCES

The archives of the Royal Society contain very full (but of course not complete) manuscript collections of relevant material. Where not otherwise noted, all manuscripts referred to in the text and notes are to be found in the Society. Most letters therein are indexed. Before December 1832 there are careful manuscript Journal Books and Council Minutes; after that date they were replaced by the printed *Proceedings* (for public distribution) and Council Minutes (for private circulation). Any MS. Council Minutes are then less complete than the printed version. Supplementary to the Council Minutes are the volumes of Miscellaneous Correspondence (cited as MC) of incoming letters, the President's Letter Book containing minutes of much, though by no means all, official out-going correspondence, and minutes of committees (CMB and DM). There is also a series of Miscellaneous Manuscripts (MM) which are bound in order of acquisition, but are calendared. There are collections of papers (mostly correspondence) of some individual Fellows, which are mostly indexed. Of these the most frequently used here are the Herschel Collection (HS) and the Lubbock Correspondence; also used was the Sabine Correspondence (Sa), partially indexed. I have not attempted to pursue the private archives of other Fellows in general, although I have used some material in the former British Museum Library, in the Wellcome Institute Library and in the Library of University College London. Much material must exist in private archives as relicts of former officers and Fellows, but I have felt that there were few problems which were not soluble with the aid of the material available to me, and that the time required to pursue all relevant archives would not have been justified by commensurate results.

Printed sources are listed in the Bibliography; as the list of abbreviations indicates, the most valuable for my purposes were the works of Weld and Lyons, in the latter case both his *History* and the earlier *Record*, which he brought up to date; the earlier editions of the *Record* were the work of various Assistant Secretaries and two Secretaries, Foster and Rücker. No other works purporting to deal with the Royal Society as a whole contain much on the nineteenth century. *Notes and Records of the Royal Society* was founded in 1939 to carry house news as well as historical articles; it is now confined to articles on the Society's history and/or the life and work of its Fellows.

NOTES TO THE TEXT

Abbreviations

BM Add. MS. Additional Manuscripts of the British Museum collection, now part of the British Library

DNB Dictionary of National Biography

DSB Dictionary of Scientific Biography

Lyons Sir Henry Lyons, *The Royal Society, 1660–1940* (Cambridge, 1944)

Phil. Mag. Philosophical Magazine

Phil. Trans. The Philosophical Transactions of the Royal Society

Procs. Proceedings of the Royal Society

Record The Record of the Royal Society of London, 4th edn (London, 1940)

Weld C. R. Weld, *A History of the Royal Society*, 2 vols. (London, 1848)

Chapter 1

1 This crisis is, of course, discussed amply in Weld, II, ch. VI, and briefly in Lyons, pp. 212–15. The original documents are chiefly collected in Royal Society Tracts nos. 3 and 4.

2 By George A. Foote in the article on Banks in the *DSB*. In 1808, when Banks was in bed with gout, he put 'The Burthen of the Royal Society Chair' on Marsden (Treasurer and Vice-President); MM 6 no. 6.

3 Ramsden, *Correspondence of Two Brothers*, p. 95.

4 HS 2, no. 22. The 'paper' (or rather letter) was on iodine and its properties. Babbage's name unfortunately does not appear in the Journal Book for 20 January 1814.

5 Barrow, *Sketches of the Royal Society and Royal Society Club*, p. 5.

6 Besides Barrow (above), see Geikie, *Annals of the Royal Society Club*.

7 It must be remembered that to sign a certificate required personal knowledge of the candidate. So on that of Robert Woodhouse, mathematical Fellow of Caius College, appears a galaxy of Cambridge names, and medical men signed the certificates of physicians and surgeons. Humphry Davy's certificate was signed by Thomas Young, but also by a number of men connected with the Royal Institution; Troughton's certificate was signed by astronomers who had used his instruments, but also by an engineer, John Rennie. When a candidate had a paper published in *Phil. Trans.* this was always noted.

8 Weld, II, p. 153.

9 Documents quoted in Weld, II, pp. 243–49, from MSS. in the Geological Society's archives. In 1902 Sir William Huggins, then P.R.S., thought that it had been of benefit to both societies to have remained separate. See especially Rudwick, 'The Foundation of the Geological Society of London', an authoritative survey.

10 In 'Circumstances of Wollaston's declining to contest the Presidency of the Royal Society with Sir H. Davy – June 1820,' quoted in L. F. Gilbert, 'The Election to the Presidency of the Royal Society in 1820'.

11 See generally Dreyer and Turner (eds.), *History of the Royal Astronomical Society 1820–1920*. For the letters of Baily and Gilbert to

Babbage, see BM Add. MS. 37182, esp. no. 237 (Baily, 11 March 1820) and nos. 205, 211 (Gilbert, 1 and 4 February 1820). Eventually William Herschel accepted the presidency on condition that nothing would be demanded of him, and remained in the office until 1823.

12 Practically endowed by the widow of William Croone, then Lady Sadleir, in 1701; the first was given in 1738 and the lectures continued more or less annually until 1830, when they were suspended until 1850 – partly because the lectureship had become a sinecure for Sir Everard Home.

13 Founded in 1774 by the will of Henry Baker to be on 'Natural History or Experimental Philosophy' and given annually since 1775; between 1794 and 1820 the lecturers were Samuel Vince (F.R.S. 1786) (seven times), Thomas Young (three times), William Hyde Wollaston (three times), Humphry Davy (six times in succession) and W. T. Brande (three times); but no name is given for the years 1814 through 1818.

14 The fund was bequeathed in 1746 by Thomas Fairchild, a botanist (never F.R.S.); increased by subscription, it paid for an annual sermon which by the nineteenth century seemed an anomalous charge for the Royal Society; hence in 1873 it was handed over to the Gardeners' Company.

15 The Copley was in Banks's later years awarded to, among others, Volta, Wollaston, Troughton, Davy, Brodie, Ivory, Brewster and the ubiquitous Home; the Rumford between 1800 and 1820 was awarded to Rumford, Malus, Davy and Brewster as well as others.

16 See, esp. Todd, *Beyond the Blaze*, based on unpublished documents including Gilbert's diary.

17 So a great deal of trouble was taken to look after the family of George Gilpin, clerk since 1785, when he died in office in 1810, as the Council Minutes show.

18 Council Minutes for 21 November 1799 and 13 March 1800.

19 Then and later the entire Council constituted the Committee, five members (from 1776 onwards seven members) being a quorum and the Committee to meet monthly or at the summons of the President. There was always provision for the Committee to call for help from non-Council members, who in this period were expected to come to the meeting, and were then allowed to vote upon *all* the papers there considered. See the Society's statutes for 1752 and 1776; revisions in 1823 and 1828 did not affect the regulations for the Committee of Papers.

20 CMB 90b covers the period 1780 to 1827, CMB 90a–f containing the complete eighteenth- and nineteenth-century records for the Committee of Papers, mostly a succinct record of author, title and action (with dates). The last 'monstrous lamb' (described by A. Carlisle, F.R.S. 1804, a distinguished surgeon) was published in 1801. Cf. variously, MC 1, nos. 37 and 38, dated 1810; and subsequent volumes of Miscellaneous Correspondence.

21 Peacock, *Life of Thomas Young*, esp. pp. 535–65. Peacock said that it was believed that Pond's mathematical knowledge and philosophical competence were too low to permit him properly to interpret his own observations, and that it was Pond's incompetence which led to the handing over of the *Nautical Almanac* to the (revised) Board of Longitude.

22 HS 2, no. 35, dated 25 October 1814.

23 HS 2, no. 98, 10 November 1818.

24 The debate was held on 6 March 1818. See Baily, *Remarks on the Present Defective State of the Nautical Almanac*, pp. 17–18; Todd, *Beyond the Blaze*, pp. 211–12, quoting Gilbert's diary; Wood, *Thomas Young*, pp. 303–5; and HS 2, no. 100 for Babbage's view (which included the belief that Pond's resignation would be forced). The astronomers wanted more astronomical data and calculations.

25 On 1 February: BM Add. MS. 37182, f. 205. George IV had very recently succeeded to the throne.

26 The original request, together with the appointment of the committee, is to be found in the Council Minutes for

11 June and 2 July 1801. The report
of the committee (which met on 12
November 1801) is in the Council
Minutes for 12 November 1801.
27 The original request in the form of a
letter to Planta (Secretary) is in the
Council Minutes for 31 March 1803.
28 Council Minutes for 3 February 1814;
for the report see Council Minutes for
24 February 1814.
29 This was an old problem. For
Franco-British cooperation in the
1670s see A. R. and M. B. Hall (eds.),
*The Correspondence of Henry
Oldenburg*, vol. VII, pp. 272, 314–16,
496–97 and vol. X, pp. 210–12. There
was in England (as earlier, and later) a
keen interest in the French
Revolutionary standardisation of
measures in nineteenth-century Great
Britain.
30 See Council Minutes for 28 March
1816, 16 May 1816 (report of the
committee meeting of 6 April) and 13
June 1816 (committee meeting of 20
May); the full report of the meeting of
1 May 1817, which was approved by
the Council for sending to the House
of Commons on 12 May 1817, is CMB
1, no. 1.
31 See CMB 1, nos. 2–4. The report is
dated 17 November 1817.
32 Council Minutes for 20 November
1817 and 8 January 1818. For the
minutes of the Pendulum Committee,
see CMB 1, nos. 2–4 (15 May 1817,
28 January 1818, 12 February 1818,
26 March 1818 and 23 April 1818).
33 Benjamin Collins Brodie (F.R.S.
1810), a pupil of Everard Home, was
already distinguished for his papers
(published in *Phil. Trans.*) on
experimental physiology. David
Brewster (F.R.S. 1815), at this period
a scientific journalist, was studying
refraction and polarisation. James
Ivory (F.R.S. 1815), professor of
mathematics at the Royal Military
College, was interested in the
application of mathematics to physical
problems, in which he employed
Continental mathematics, and already
had a European as well as British
reputation. Robert Seppings (F.R.S.
1814) was an innovative naval
architect. Brodie and Seppings were
already Fellows when they received

the Copley Medal but Brewster (1815)
and Ivory (1814) were not.

Chapter 2

1 The whole story of the succession in
1820 is told very fully with extensive
use of MS. sources by L. F. Gilbert,
'The Election to the Presidency of the
Royal Society in 1820' (hereafter
L. F. Gilbert), where, however, no
precise MS. references are to be
found.
2 In his diary for 8 May and 25 April,
quoted in Todd, *Beyond the Blaze*, pp.
213–14.
3 According to Gilbert's entry in his
diary for 1819, Banks had decided to
appoint him Treasurer, in succession
to Lysons after the latter's death, and
then Vice-President, all a month
before he was 'elected' to the Council.
Todd, *Beyond the Blaze*, p. 213.
4 Council Minutes for 1 June 1820 and
Todd, *Beyond the Blaze*, p. 214.
5 The story was retailed to the Duke of
Somerset by Sir Alexander Johnstone,
who remarked 'This story has as you
may suppose not added to Home's
popularity'. See Ramsden,
Correspondence of Two Brothers, p.
275, also quoted by L. F. Gilbert, pp.
270–71. Babbage told Herschel of
being canvassed on Prince Leopold's
behalf immediately after Banks's
resignation (HS 2, no. 136 of 22 May,
also quoted by L. F. Gilbert, p. 269)
by his physician Dr John M'Culloch
(F.R.S. 1820) who reported that the
Prince was determined to 'devote
himself' to the duties of the office if
elected.
6 See above, Chapter 1, p. 6 and its
note 11.
7 Quoted in Cameron, *Sir Joseph Banks*,
p. 114; the letter was printed in the
issue for 24 June 1820. Cf. The view
of John Barrow: 'In truth the chair of
the Royal Society does not require to
be perpetually and exclusively filled by
men of science, or by persons elevated
in any one particular department of
science. The President should be
conversant in general knowledge,
especially in the knowledge of the
world' (*Sketches of the Royal Society
and Royal Society Club*, p. 37). For a

rather different point of view, more
favourable to Davy and his subsequent
Presidency, see Miller, 'Between
Hostile Camps: Sir Humphry Davy's
Presidency of the Royal Society of
London, 1820–1827'.

8 In his personal account,
'Circumstances of Wollaston's
declining to contest the Presidency of
the R. Socy with Sir H. Davy. – June
1820 –' quoted in L. F. Gilbert, p.
259. Davy had been thought
ungenerous to Berzelius on his visits
to England; he certainly was to be so
to Faraday. Miller (note 7 above)
regards Davy as, on the contrary,
representative of the scientific
movement against the 'old guard', but
the situation is far from clear and, as
Miller notes, Davy was on the whole
an ineffective President.

9 This was a point that Babbage was to
reiterate in 1830, entitling a section of
his *Reflections on the Decline of Science
in England* (pp. 188–90), 'Of the
Influence of the Royal Institution on
the Royal Society'.

10 One survives, printed in Granville
(ed.), *Autobiography of A. B. Granville*,
p. 187, also reproduced in
L. F. Gilbert, pp. 263–64. Granville
(F.R.S. 1817) was a physician, so
Herschel brought in the name of Dr
John Latham (F.R.S. 1801), as it
happens an Oxford graduate, who was
a warm supporter of Wollaston. But
Davy had been in some sense
Granville's patron, and they were by
now friends, so Granville supported
him and canvassed on his behalf.

11 Cf., e.g., the opinion of Leonard
Horner (geologist, F.R.S. 1813) in a
letter to Marcet of 8 November 1820:
'I had the mortification of hearing
some details of the new Presidency of
the R.S., and that if Wollaston had
stood firm he had got a triumphant
majority' (Lyell, *Memoir of Leonard
Horner*, I, 176).

12 Sir Alexander Johnstone reported to
Somerset apropos of the Royal Society
Club dinner of 6 July 1820, 'Davy put
on a meek modest appearance having
been informed by his friends that the
manner in which he had interfered
with the proceedings of the R. Society
had caused disgust, & Wollaston

presided with the diffidence suitable to
a temporary president' (Ramsden,
Correspondence of Two Brothers, pp.
275–76, quoted in L. F. Gilbert, p.
270). Dr John Bostock (F.R.S. 1818,
chemist, physiologist and geologist, in
1832 a Vice-President of the Royal
Society) told Marcet that Wollaston
was 'less generally popular than his
good qualities & great talents should
render him; people talk much of his
reserve & distance of manner, & this
so much so, that we must suppose he
has a considerable share of reserve in
ordinary society, tho' not among his
more intimate associates' (quoted by
L. F. Gilbert, p. 266). For his manner
of discussing scientific differences with
equals, see his letters to Young, MS.
Young (242) nos. 44–51, esp. no. 48
(1800–1801); but in 1821–23 Faraday
was to find him decidedly prickly (see
Jones, *Life and Letters of Faraday*, I,
pp. 299–313), although in the end
Wollaston became friendly and
supported his election as F.R.S.

13 Davy to Thomas Poole, 10 December
1820 in Paris, *Life of Sir Humphry
Davy*, pp. 370–71, partly quoted by
L. F. Gilbert, p. 268. Davies Gilbert
has been one of Davy's early patrons.

14 BM Add. MS. 37182, f. 272. This was
in effect a canvassing letter. The
President was properly elected from
among the Council, but election of
Council members preceded the vote
for the President.

15 Ramsden, *Correspondence of Two
Brothers*, pp. 277–78 (8 December
1820). Johnstone cited Wollaston's
'paper' about the award of the Copley
Medal to Oersted in most favourable
terms, declaring 'The style &
arrangement of the matter...does
Woolaston [*sic*] credit – & affords in
my opinion a good model for writings
upon philosophical subjects'.

16 *Ibid*. Davy published this speech as
the first of *Six Discourses* in 1827, and
by so doing stirred up much criticism,
especially from Babbage, because he
appeared to have published the
collection of his Anniversary addresses
at the Society's expense.

17 Lyell, *Memoir of Leonard Horner* (21
February and 10 April 1821), I, pp.
185 and 191.

18 Paris, *Life of Humphry Davy*, p. 370. Paris himself (F.R.S. 1821) was doubtful that a scientist was in fact the best man for P.R.S.

19 For example, the editors of the *Phil. Mag.* wrote in July 1820 (**56**, p. 58) 'Sir Humphry Davy is expected to be the new President [of the Royal Society]. The Society could not make a choice more acceptable to the friends of science'. Miller (note 7 above) gives other instances of support, and writes at some length about Davy's influence for reform.

20 Babbage was to claim later that when the Secretaryship was vacant in 1826 Wollaston and others on the Council wanted to appoint him to fill it, but that Davy, after at first agreeing, later chose Children, claiming that the President had the right of nomination. (Children was probably the better man for the job.) Children was replaced by Sabine in 1827, but appointed again 1830–37. Cf. Babbage, *Passages from the Life of a Philosopher*, pp. 186–87.

21 See Council Minutes for 2, 15 and 23 June 1825. The Royal Society preferred the form 'Royal Literary Society' (giving F.R.L.S.).

22 The Committee consisted of the officers, Daniel Moore (F.R.S. 1810, Council member 1820–21), Thomas Murdoch (F.R.S. 1805, Council member 1813–17, 1822–23), Wollaston and Young (Council Minutes for 17 April 1823). The reforms limited the number of Foreign Members (but this was never a problem) and changed the post of 'Clerk' to that of 'Assistant Secretary'.

23 Whewell to Herschel, 15 October 1823, HS 18, no. 163. The paper was read on 25 November 1824 and published in *Phil. Trans.*, **115**, 1825, 87–130.

24 Buckland Correspondence, no. 10 (18 March 1822). On 21 November (no. 14) he wrote of the Council's having just voted Buckland the Copley Medal for his paper on fossil teeth, printed in *Phil. Trans.*, 1822; it had been read on 22 February.

25 The earlier paper 'on the form of Cogs, either capable of rolling without friction, or of imparting equal angular velocities' was read on 8 January

1807, with the introductory statement 'In compliance with a request of the President Mr. Giddy [Gilbert] makes the present communication, but expresses a wish that Professor Farish of Cambridge, who had arrived at the same conclusion nearly at the same time with himself could have been induced to communicate to the Society the steps of his investigation'. Gilbert's paper was 'postponed' when the Committee of Papers considered it on 19 March 1807; he withdrew it on 14 May 1807. In 1826 he was anxious and uncertain that his paper would be printed after its reading on 9 March 1825 (see HS 8, no. 109), although he was considered an authority on bridge design, being successively consulted about Telford's Menai Bridge (completed 1826) and the Clifton Suspension Bridge. See Todd, *Beyond the Blaze*, pp. 99–100.

26 The Committee, which included South (an instigator), Beaufort and Herschel, drew up a full report on 19 April 1827; see CMB 1.

27 As revealed by letters in MC 1.

28 Todd, *Beyond the Blaze*, p. 223 (Gilbert's diary for 28 March 1822).

29 Cf. HS 2, nos. 207 and 208 (Babbage and Herschel letters of December 1826 and 30 January 1827). Herschel thought South had behaved badly in attacking Babbage's proposal of the candidature of Cesar Moreau on trivial points, but also that Babbage had over-reacted, especially in denouncing South's candidate, and that the whole affair had been 'bad for the credit of the Royal Society', adding, 'it is not the course to begin reform by the degradation of the body to be reformed'. In the event, Babbage's candidate was elected and South's was not.

30 See CMB 1, p. 42 *et seq*. This is the catalogue which Panizzi found defective (see below, Chapter 3). Regrettably, indexing of the Council Minutes soon lapsed.

31 Council Minutes for 24 November 1825. South was allowed indefinite loan. It was apparently in the 1820s that Leeuwenhoek's microscopes disappeared.

32 Council Minutes for 6 April 1826; see

also Whewell's letters, HS 18, no. 165
et seq., where he gives a lively account
of the experiments which they had to
abandon when the instruments were
dropped nearly the whole depth of the
mine.
33 Minutes in CMB 1 *passim*. See esp.
under 14 March 1822 and 12
December 1822. The Committee met
until 1826.
34 Peel's letter to the Society of 15
December is printed in Weld, II, p.
401. See also his letter to Davy of 13
December 1824, and that to George
IV proposing the medals in Parker,
Sir Robert Peel, I, pp. 363–64 and
387.
35 See *Passages from the Life of a
Philosopher*, pp. 188–89 and *The
Decline of Science*, pp. 115–24. The
outside world seems to have heard of
no criterion except 'worth': cf. *Phil.
Mag.*, **66**, 1825, 461.
36 HS 10, no. 255 (November 1828).
Herschel in reply (28 November)
reassured Ivory: 'In decreeing you
one of their Royal Medals they
honoured themselves in rewarding
distinguished merit'. Ivory was
elected a Corresponding Member of
the French Institut in 1828.
37 For the Treasury's request, see
Council Minutes for 17 April 1823.
The committee (the officers, Baily,
Brunel, Colby (of the Ordnance
Survey), Gilbert, Herschel, Kater,
Pond, Wollaston and Young) gave the
first report at the Council Meeting of
1 May 1823 (also in CMB 1).
38 See HS 18, no. 323 (4 April 1822); in
no. 324 (n.d., but almost certainly
earlier) Young, after asking Herschel
to check data in 'the Nautical
Almanac for the 20 Aug. 1819' added
'Many thanks for your note: but pray
do not omit to tell Mr. S. that he is
bound to give his *reasons* – & the
sooner the better'. Baily's attacks were
(1) Preface to *Astronomical Tables and
Remarks for the year 1822*, to which
Young replied in defence, an answer
printed by Baily in (2) *Remarks on the
Present Defective State of the Nautical
Almanac* (also 1822). For other
criticism, see *Phil. Mag.*, **50**, 1818,
186–87, **51**, 1818, 146–47, **53**, 1819,
217–19 and **60**, 1822, 325–26.

39 Council Minutes for 26 February
1824; at the next meeting (11 March)
it was decided that there ought to be
tables of 'Precession, Aberration,
Solar Nutation & proper motion for
the principal Stars for every day in the
period of four years including leap
year and that a separate Table should
be given for the lunar nutation to
every degree of the Moon's Mode'.
Young must have wondered how he
would get assistants; as he had pointed
out to Herschel earlier, skilled
astronomers in England did not want
to spend their time in the drudgery of
calculation when they could be doing
original work.
40 Lee's letter was noted at the Council
Meeting of 3 March 1825 and
discussed at length with Pond's reply
on 5 May. The committee then
appointed (Baily, Colby, Gilbert,
Herschel, Kater, Wollaston and
Young) reported on 16 June. In
November Lee wrote to Croker, as
one of the Secretaries to the Admiralty
(who, however, could not see what it
had to do with him), as noted in the
Council Minutes for 17 November
1825; on the 24th Lee was asked to
apologise or explain his conduct in
writing to Croker, which the Council
thought 'highly improper &
indecorous'. His reply, read on 15
December, was so unsatisfactory that
it was decided not to enter it into the
minutes 'in consequence of the
offensive expressions which it
contains'. Resignation was clearly
Lee's only recourse.
41 See Council Minutes for 6 April 1826
(with prolonged correspondence with
Pond), 18 December 1823 and 18
November 1824, respectively.
42 See Council Minutes for 13 February
1823 and 24 June 1824; for Davy's
activities see Paris, *Life of Humphry
Davy*, pp. 399ff. and *DSB* entry for
Davy.
43 The Optical Glass Committee minutes
for 1824–28 are in CMB 1. For
Young's letter on behalf of the Board,
see Thompson, *Michael Faraday his
Life and Work*, p. 97, and (for
Faraday's conclusions) p. 99, as also
his Bakerian Lecture for 1829 (*Phil.
Trans.*, 1830).

44 Paris, *Life of Humphry Davy*, pp. 448–49.

45 See Gillen, *Royal Duke*, pp. 182–83, quoting letters in the Royal Archives, and, for Gilbert's private views, Todd, *Beyond the Blaze*, pp. 221–39 which covers the whole 1827 crisis in an interpretation somewhat different from that given here.

46 Gilbert's letters to Babbage are BM Add. MS. 37183: for 1823, see esp. ff. 1–2 and ff. 25 *et seq*. For his steam engine work see Todd, *Beyond the Blaze*, esp. pp. 57–111, and *Phil. Trans.*, 1830, pp. 12ff. for his paper 'On the Progressive Improvements made in the Efficiency of Steam engines in Cornwall', which includes his (original) definition of 'duty'.

47 See Todd, *Beyond the Blaze*, p. 224, quoting Gilbert's diary. For the subsequent negotiations see BM Add. MS. 40394 (Peel Papers), ff. 235–86, which contains letters from Gilbert, Amyot, Barrow and others about the day-to-day 'state of the poll' during November 1827.

48 For Wollaston's view, see his letter of 26 November [1827] to an unnamed correspondent, quoted by L. F. Gilbert. For Gilbert's viewing himself as a 'caretaker' President paving the way for Sussex, see Todd, *Beyond the Blaze*, p. 239.

49 The minutes are carelessly entered in the form 'that the Papers, proposed,...&c. to be considered by the Committee of Papers...'. Since the Committee of Papers consisted of the whole Council I have assumed that the minutes should read 'proposed to be read to the Society, to be considered by the Sub-Committee of Papers...'.

50 As Peacock (F.R.S. 1818, from 1839 Dean of Ely) declared in 1834 to be the case with many of his Cambridge contemporaries (see letter to Lubbock 7 May, Lubbock Correspondence 30, P 101 bis).

51 See Lyons, pp. 234–41.

52 Wollaston's letter of 26 November 1828, witnessed by Warburton, who promptly gave a hundred guineas to the fund, is printed in Weld, II, pp. 447–48, with the list of immediate subsequent gifts (including £1000 from Gilbert); see also Council

Minutes for 11 December 1828. Young's speech, reported by himself, together with his shyness, is quoted in Wood, *Thomas Young*, p. 327. Sadly Wollaston died at the end of the month.

53 Gillen, *Royal Duke*, p. 182.

54 Todd, *Beyond the Blaze*, p. 242. Gilbert recorded in his diary for 1 December that, being approached through Pettigrew 'I gave a civil answer enclosing a Paper to be shewn to his R.H. in which I assigned my reasons for thinking it inexpedient for the Royal Society to have a President from among the King's near Relations and therefore declining a compliance with his request'. But he clearly did not think this himself, or not often.

55 Young to Herschel, HS 18, no. 328 (n.d.), in reply apparently to Herschel's no. 330, and also Young's no. 329, where he says 'It is true he does not want a medal – but the medal wants the honour of being given to him'.

56 To Herschel, 24 August 1828, HS 18, no. 344. Parliamentary details from J. H. Barrow (ed.), *The Mirror of Parliament* for the second session of the 8th Parliament...II (London, 1828), s.v. 26, 27 June, 1, 3, 4 July.

57 On 17 July 1828, HS 21, no. 19.

58 See Council Minutes for 22 January 1829 and MC 1.

59 Letter to the Admiralty 10 January 1829, quoted in Wood, *Thomas Young*, p. 311. Later, in a letter to a friend (*ibid.*, p. 312) he remarked 'There are certainly some excuses for the affair of the Board of Longitude in the encouragement which it held out to the ravings of madmen' – that is, those with impractical solutions to the problem of determining longitude at sea. Baily's criticisms are to be found in his *Further remarks on the Present Defective State of the Nautical Almanac for 1822* and South's in *Refutation of the numerous mis-statements and fallacies contained in a paper presented to the Admiralty by Dr. Thomas Young* (25 April 1829).

60 *The Lancet*, 13 and 20 June 1829, respectively. It is not perfectly clear on which side 'H' was over the Anatomy Bill, which Wakley strongly favoured.

61 But possibly also directed rather against Davy, for two years later he was to write a different view to Daubeny, then professor of chemistry at Oxford: 'I see you have been cited by Brewster & his Coadjutors as authority on the "Decline of Chemistry". Question – I think they have carried the joke too far – we have *good* chemists – witness yourself – Turner – Faraday – & the last named especially has I think been handled in a very unjustifiable way by B' (HS 21, no. 102). And in 1828 (21 July, HS 21, no. 20) he wrote to Faraday to encourage him to continue his glass experiments, although he expected the Royal Society's Optical Glass Committee to vanish with the abolition of the Board of Longitude.

62 HS 18, no. 347 of 1 September 1828.

63 According to a letter of Young to a friend in December 1828 (Wood, *Thomas Young*, p. 327) South 'proposes that the Society at large should supersede me in the appointment of Foreign Secretary which has hitherto been a "job" of the President and Council and should make Mr. Ivory my successor, whose qualifications are a total ignorance of all foreign languages, and a disposition to quarrel with all persons of merit, almost as great as that of Mr. South himself'. Young, a good linguist, was a conscientious Foreign Secretary. The post itself had only been formally added to the list of officers in the revised statutes of 1823.

64 The dedication, perhaps oddly in the circumstances, to 'a Nobleman whose exertions in promoting every object that can advance science reflect lustre upon his rank' is dated 29 April 1830. It is often said that this was Humboldt, but there is no solid evidence. Babbage had been working on the book for two or three months and had essentially finished it by mid-March (HS 2, no. 246). His motives in writing are stated forthrightly in *Passages from the Life of a Philosopher*, esp. pp. 86–87, 144–46. It should be noted that Babbage was never elected to the Institut. Young was (1818/1827).

65 He says, with approval, that Wollaston

had suggested limiting numbers to 400 (there being then over 700 Fellows). He himself preferred the idea of distinguishing the Fellows into two classes: those who had, and had not, contributed to *Phil. Trans.* – taking no account of any other scientific contributions. For further details of Babbage's failure to achieve the Secretaryship see Miller (note 7 above), p. 39.

66 In writing up the minutes for 26 November 1829 Roget took an understandable short cut in recording that the Council had put forward Sir John Franklin as a Council member for 1829–30, when in fact it had voted for Francis Beaufort (the hydrographer). According to Roget and Gilbert, the latter told Roget immediately after the meeting that Beaufort had already declined, so Roget silently omitted his name in favour of Franklin's, the latter having been discussed although outvoted by Beaufort. Beaufort was known to be in favour of reform.

67 HS 2, no. 246, 19 March 1830. In fact his anger never subsided and he was to recur to his charges in *The Exposition of 1851* (1851) and his autobiographical *Passages* of 1864.

68 Thus when Lubbock (who agreed with Babbage as regards *London* science) asked whether Babbage would mind meeting Sabine at dinner, Babbage replied loftily that there was nothing personal about his attack on Sabine, although he saw that the latter might not understand 'the feelings of men of science' (BM Add. MS. 37185, f. 130, 23 April 1830). (Evidently Lubbock was well aware of the content of Babbage's book.) Sabine attempted a refutation in *Phil. Mag.*, n.s., 8, 1830, 44–50. The Duke of Somerset elicited a similar reply when he wrote doubting the propriety of having invited Gilbert and Babbage to dinner together (BM Add. MS. 37185, f. 170).

69 Herschel's first (undated) letter is HS 2, no. 245, and the second, of 22 May 1830, no. 252. He correctly saw that Babbage's views might be taken to be those of the Astronomical Society.

70 Brewster told Babbage that he had

intended to write about the decline of science in England himself (BM Add. MS. 37185, ff. 49–50). For his review of Babbage, see *Quarterly Review*, **46**, 1830, 305–42. Lubbock told Babbage that he himself thought that it was only in London that science was in decline (BM Add. MS. 37185, f. 139). For Dalton's letter, see *ibid.*, f. 176 (dated 15 May 1830). J. F. Daniell devoted most of his Inaugural Lecture at King's College, London (1831) to the topic, pointing out that London was rising even if the older universities were somnolent. Moll's *On the Alleged Decline of Science in England*, although only thirty-three pages long, is a most effective counter to Babbage, and managed to point out that Babbage had already held the Lucasian Professorship at Cambridge for four years (he was to hold it for eight more) without lecturing or even residing in Cambridge. Faraday always agreed with Moll that scientists should not require honours to do good work (see Williams, *Michael Faraday*, pp. 352–53). Brewster attacked both Moll and Faraday in *Edinburgh Journal of Science*, **5**, 1831, 343ff.

71 For his friends' letters, see BM Add. MS. 37185, ff. 185 (South), 194 (Baily) and 197 (Capt. George Everest); for Babbage's replies to Gilbert, see ff. 200–203. Roget (f. 189) had told him some Fellows had proposed that Babbage should be summoned to answer personally.

72 *Phil. Mag.*, n.s., **7**, 1830, 446–48, **8**, 1830, 72–73, 153–54 for Roget, Gilbert and Babbage, letters reprinted in *The Times* for 26 June 1830. (Cf. Williams, 'The Royal Society and the Founding of the British Association for the Advancement of Science'.) For criticism of Babbage see the letter by 'Socius' in *Phil. Mag.*, n.s., **8**, 1830, 354–55. For the facts see note 66 above.

73 Kater had written to Babbage the previous week that he felt it his duty to the Society to ask the Council about possible action, but wanted Babbage to know that he remained personally friendly (BM Add. MS. 37185, f. 210). This was generous, in view of Babbage's published attacks on army

officers active in scientific affairs (Kater was an engineering officer).

74 BM Add. MS. 37185, ff. 250 and 254. This is in line with contemporary summing up of Gilbert's character.

75 BM Add. MS. 37185, f. 252, South to Babbage, 8 July 1830.

76 The following account draws heavily on the anonymous pamphlet *A Statement of Circumstances Connected with the late Election for the Presidency* (London, 1831), usually ascribed to W. H. Fitton (F.R.S. 1815, on the Council 1827–30, M.D., F.C.P., secretary (1822–24) and president (1827–29) of the Geological Society) who was known in January 1831 to be preparing a work which Whewell and others thought 'tame' (R. Jones to Herschel, HS 10, no. 350). This gives a running chronology and quotes Gilbert's letters to Pettigrew, 4 September–5 October, with abstracts of Pettigrew's replies, as well as extracts from the *London Literary Gazette* and from the Council Minutes in a surprisingly impartial manner. The Council Minutes for the period are adequate but naturally brief. Letters, now in the Royal Archives, exchanged between Pettigrew and the Duke of Sussex are quoted in Gillen, *Royal Duke*, pp. 182–83. Fitton opposed Sussex but was friendly with Gilbert, and letters, now lost, were, according to Todd, exchanged in October 1830 (*Beyond the Blaze*, p. 253, note 2).

77 MM 5, no. 34, dated 18 September 1830.

78 The first edition dated 11 November 1830 was followed by a second dated 25 November 1830, the latter adding the complaint that the copy given to the Society by South on 18 November was neither listed as a 'present' nor placed in the Library. Copies of both editions, annotated by South, are now there (Royal Society Tracts RS/4 and RS/5).

79 Gilbert to Herschel, 29 June 1830 (Todd, *Beyond the Blaze*, p. 251) and Herschel to Gilbert, 1 July 1830 (HS 8, no. 127). For earlier correspondence, in which Gilbert tried to persuade Herschel not to resign from the Council in 1829, see HS 8,

no. 120 bis and no. 122. Babbage's *Ninth Bridgewater Treatise* went through two editions in 1837 and 1838.

80 Herschel to Fitton, 18 October 1830, quoted in Todd, *Beyond the Blaze*, p. 254.

81 Reproduced in Fitton's *Statement*; in the Royal Society's copy the broadsheet has been tipped in. According to Fitton, the sixty-three signatories were soon joined by seventeen more. For Fitton's activities in the campaign, see further Morrell and Thackray, *Gentlemen of Science*, pp. 54–56.

82 On 26 November 1830, HS 2, no. 257. He was, he said, glad to see that Babbage's name was not on the list, which made the affair seem less '*violent*'.

83 *Science without a Head*, p. 102.

84 Cf. George Harvey to Babbage, 3 January 1831 (BM Add. MS. 27183, f. 429); he professed to have had his coat on when he received Babbage's note saying that it was unnecessary for him to travel to London to vote. It is possible that Herschel was identified by some with 'a Cambridge faction'.

85 Letter of 21 December 1830, HS 12, no. 383. Clearly Murchison was guilty of exaggeration.

86 *The Lancet*, 4 December 1830, p. 338, leading article, corrected by a correspondent 18 December, p. 403, and a letter signed 'Zero', 25 December 1830.

Chapter 3

1 The sources for the following reconstruction, besides the Council Minutes, the Journal Book and, after 1832, the *Proceedings*, are mainly to be found in MC 1, no. 259 *et seqq.*, the President's Letter Book (MS 425) which begins in December 1830, and the Lubbock Correspondence, especially C 112 *et seqq.*, letters of Children, and P 99 *et seqq.*, letters of Peacock.

2 Lubbock Correspondence, C 224, 2 November 1835, quoting a letter written by the President.

3 *Ibid.*, C 160, 1 November 1832. But he gave good reasons, intellectual, not personal, for his choices, aiming at a balanced Council, including men with political opinions opposed to his own, as appears from C 154 of 17 September 1832.

4 C 183 of 9 September 1833.

5 The letter is in the printed Council Minutes for 14 September 1838, I, pp. 182–84; the original is in MC 2, no. 297. Peacock was apparently influential in persuading him to resign (Wellcome Institute MS. 63700).

6 This matter was so troublesome that Children compiled a separate Evans MS. (MS. 122) of the whole affair, including the letter to the candidate which began the difficulty, those exchanged by Children with members of the Council (his 'unfortunate words' seem to have made virtually no impression on most of them) and a narrative account drawn up after Children's candid interview with the Duke of Sussex. It should be noted that it was expected that a candidate's character should be unimpeachable, and Evans was by no means the only person to be blackballed on account of financial irregularities, as also that Sussex's rashly slanderous remark to Pettigrew was unknown to the Council when voting.

7 Lubbock Correspondence, C 126, 19 April 1832.

8 See *ibid.*, esp. C 112 and 113 of 28 and 30 December 1830, and P no. 99 of 21 December 1830. As it was to turn out, the Treasurer (Lubbock and then Baily) normally presided in the President's absence.

9 The system of refereeing was first announced on 30 November 1832 (*Procs.*, no. 11). Marshall Hall's complaint and appeal to the Council on 22 June 1837 was typical; in this case he possibly was unfairly treated, for Roget was not to show himself always impartial. See MC 2, nos. 255, 272, 273 and 279, and HS 9, no. 198, Hall's appeal to Herschel of 10 October 1839, where he complains that Roget's abstract of his paper (published in vol. III of the abstracts) was unfair, that no one would take the trouble to test his work experimentally, and that, not being on the Council, he could never get his

paper printed and so could not be in the running for a Royal Medal. Hall was extravagant, as Herschel clearly thought, and as Hall's printed 'Letter addressed to the Earl of Rosse, President-elect of the Royal Society' (Royal Society Tracts 292) shows, and persistent in his grievance, but as was to appear Roget's conduct was unwise, to say the least. For what follows see Council Minutes for 8 February and 8 March 1838.

10 See Council Minutes for 7 January, 14 April, 8 and 23 June 1836, the Journal Book (not *Procs.*) for 28 April and 18 June 1836, and MC 2, nos. 198, 205–209. It is not clear quite why this ruling was necessary.

11 The minutes of the Committee are in CMB 30. It met on 7 and 16 May, and 14 July, with the President presiding at the first and last meetings. There was also a sub-committee; for its report on 30 June see DM 1, no. 38. For the refusals of various Fellows (these include Herschel, Whewell, Petit, Baily, Fitton, R. Brown, Babbage and Penn), see DM 1, and also Council Minutes for 17 March 1831. Most either claimed to be too busy or politely declined; only Babbage was rude, declaring 'having no reason to imagine that my presence on that Committee would be attended with any advantage to science I beg leave to decline the nomination'. Babbage pursued a policy of non-collaboration even in regard to his own 'engine', telling Herschel that 'it would be degrading to act on their Committees' (HS 2, no. 259). South had already refused to serve on the Optical Glass Committee, on the grounds that, as he claimed, Royal Society committees often reported as opinion what no one on the committee had said (MC 1, no. 265).

12 See in particular the report by the Society's solicitor (Few) of 4 May 1831 in the President's Letter Book.

13 Although finance is clearly important it makes dull reading and consequently no details will be given here. They are available at considerable length in Lyons, ch. VII and ch. VIII, *passim*.

14 For all this, besides the Council Minutes (which include Panizzi's letters) there are the minutes of the Catalogue Committee established in 1832, CMB 47a and 47b, and in Royal Society Tracts 6 are to be found Panizzi's printed letters (nos. 4, 5 and 7), the Committee's *Defence* (no. 6) and the Presidential addresses of 1836 and 1837 (no. 8). Ironically, when the post (a part-time one) of Librarian to the Society was vacated in 1835 on Roberton's becoming Assistant Secretary, Panizzi applied for it, but Shuckard, who subsequently catalogued the Society's manuscripts, was in fact appointed.

15 See Lubbock Correspondence, L 443, 445, 447, 448 and R 126, 129. The word used was 'tyranny' as Roget told Lubbock. Lubbock wrote to Gilbert, because he had originally appointed him Treasurer.

16 Anonymous review of Barrow's *Sir Joseph Banks and the Royal Society* in Sharpe's *London Magazine* for 1845 (pp. 221–35, 351–69, 565–86); this is attributed in the Society's copy (Royal Society Tracts 280/1) to Fitton, but it seems too acidulous for him. The author complains that 'Social status, or at least opulence are, in a great degree, essential to a Fellowship' and claims that scientific candidates were often blackballed.

17 See Morrell and Thackray, *Gentlemen of Science, passim*, esp. p. 9, quoting Northampton's speech at the 1837 Liverpool Meeting of the British Association. For Burlington, and Murchison's ideas, see *ibid.*, pp. 55 and 436–37 (the latter is Wellcome Institute MS. 63700, Murchison letters, 29 October 1838). Herschel settled for a baronetcy secured for him by the Duke of Sussex: see HS 2, no. 290. For Hamilton's opinion see Lubbock Correspondence, H, no. 47.

18 See especially his letters to Lubbock 1838–45, Lubbock Correspondence, 10, C, 417ff.

19 Letter of 22 February 1848, MC 4, no. 217.

20 It was more thoroughly reviewed in 1842: report of committee appointed 10 March to the Council on 14 April. It is printed there, and subsequently

21 Walter White, *Journals* (see below and note 27); he was at this time assistant to the Assistant Secretary. Mantell's *Journal* (ed. E. C. Curwen) is less informative.

separates were run off for 'occasional distribution'. A further committee was appointed 9 June.

22 MC 3, no. 310 is Shuckard's letter of 9 November 1843 in reply to a query by the President at the previous Council meeting, assuring the Council that he had passed unopposed through the Insolvency Court. He attributed his financial difficulties to his addiction to entomology.

23 See Lyons, pp. 259–62, relying mainly on Gassiot's 1870 pamphlet, *Remarks on the Resignation of Sir Edward Sabine K.C.B. of the Presidency of the Royal Society*, Royal Society Tracts 695, no. 8. The Charter Committee minutes are DM 1, nos. 40–60.

24 See the letter of Sir John Richardson, naval surgeon and arctic explorer, Council Minutes for 17 December 1846.

25 See the Minute Book, and Bonney, *Annals of the Philosophical Club*.

26 The following account is based on the minutes of the Committee of Physiology 1838–49; printed Council Minutes 1845–47; MC 4, nos. 94, 98, 102, 108, 116, 125, 126, 173, 196, 310; President's Letter Book 1846–48 *passim*; *The Athenaeum* 1847; *The Lancet* 1846–48, which includes an apparently verbatim report of the Special General Meeting on 11 February 1847. The Referees' Reports (MS. RR 1) 1832–49 show careful refereeing in general, especially for Lee, but contain only the general report on Beck.

27 *Journals of Walter White* for 22 November 1845: 'The Committee which awarded the Royal Medal to Mr. Beck are to reconsider the grounds of their decision.' See also his record (2 December 1864) of a conversation with 'Dr. Gray' (J. E. Grey, naturalist, F.R.S. 1832, who was on the Committee in 1845 and 1846) in which he said the award to Beck 'was a job; that Dr. Todd and Mr. Bowman were determined Dr. Lee should not have a medal,...that

Dr. Roget...laid before the Committee a false minute', and that he [Grey] had written a long account to the President asking him to lay it before the Society, to which Northampton agreed, but said that if so he must resign 'on which he (Dr. G) exonerated him'.

28 The letter by Jones is MC 4, no. 98. Lee's letter dated 1 December 1845 is no. 102; see also nos. 108, 116, 125, 126 in 1846. A paper by Lee was part of the list considered, but as its title 'Supplement to his paper on the nervous ganglia of the uterus' indicates, his main paper on the subject was earlier, although still eligible. It is perhaps worth noting that there is a considerable entry for Lee in the *DNB*, but none for Jones or Beck. Ironically, in 1847 the Polish-German nervous physiologist Robert Remak was to complain that neither Beck nor Lee acknowledged *his* work; see MC 4, nos. 173 and 310 (Lee's reply). Lee reprinted all his letters (including those of 22, 30, 31 December 1847 and 4 January 1848, with replies by Northampton and Christie, and much else in the fourth of his *Memoirs on the Ganglia and Nerves of the Uterus*.

29 President's Letter Book, Weld to Lee 10 February 1846 (and 18 February). It was a feeble reply, not only because precedence of illegality is a poor defence, but because the recipients to whom it applied (notably Lyell) had published elsewhere, as Beck had not done, although publication was announced. Christie was of course not directly involved in the award, as Roget had been.

30 Leader, 21 March 1846. Wakley was later to claim that Roget's Bridgewater Treatise was based on Grant's lectures, and that Roget bore Grant a grudge because Grant would not proofread for Roget (4 April 1846, pp. 391–93 and 11 April, pp. 418–20, where Newport, as Grant's pupil, was said to have also been asked to proofread); Grant himself claimed Roget to have been an assiduous note-taker at his lectures (18 April, pp. 445–46). Newport, a virtually self-taught entomologist (F.R.S.

1846), was to receive a Royal Medal in 1851.

31 The Society continued to observe its seventeenth-century rule of not judging between controversialists, and in the nineteenth century always insisted that papers should be complete in themselves without the need for experimental demonstration outside the Society's premises. Old Sarum was one of the most notorious rotten boroughs, being in the nineteenth century as now completely deserted.

32 Hall's letter to the editor of *The Lancet* is in the issue for 18 April 1846, p. 451; Wakley's leader is in the next issue, 25 April, pp. 468–74, quoting letters between Children and Hall in 1837 over a paper of Hall's which was felt to require revision and which Hall finally reluctantly withdrew, blaming Roget. Hall also complained that no one would witness his experiments. See also Hall's final version 'A Letter addressed to the Earl of Rosse, President-elect of the Royal Society' [Dated 15 November 1848], of which there were apparently two 'editions' (Royal Society Tracts 292), where he complains further of not having been on the Council nor the Committee of Physiology nor been asked to give a Croonian Lecture (in abeyance 1830 to 1850) or to be a referee. In view of his temperament the last is understandable. He felt Roget had injured him, knew it, and acted accordingly. It is true that Roget was not present when the Committee of Physiology voted to propose Hall's name for the Copley Medal. Hall believed that there should be no 'secret and anonymous Reports' on papers, and that there should be more changes in the Council and that experiments should be witnessed.

33 Issues for 30 May, 6 and 13 June 1846. In July there was a complaint from a Weymouth surgeon that his paper was read and not published, and that Owen's published paper on the same subject did not mention him.

34 Issue for 16 January, leader, p. 74; cf. the leader for 23 January, p. 101: it is 'not merely one act, but a long career, of injustice with which the Royal

Society has been charged'. The very full account is in the issue for 20 February 1847, pp. 191–96. A less full and less unfavourable account is in *The Athenaeum*, 13 February 1847, pp. 174–75. According to White, Northampton spoke 'with much tact and temper. Altogether the discussion was characterised by great fairness, no attempt at quibble or concealment. A forest of hands was held up in favour of the Council, *three* only for the opposition.'

35 This account may be supplemented by Curwen, *Journal of Gideon Mantell*, entries for November 1849.

36 The letter itself (not printed in full in the Council Minutes) is MC 4, no. 217.

37 An anonymous motion for a limitation to five years was made on 1 November 1838 and immediately negatived; on 10 March 1842 Fitton had notified the Council that he intended to propose a limitation of two years, while de la Beche proposed four years (24 February 1848); these proposals were not pursued. The subject arose again only during Sabine's Presidency.

38 See Lyell to Herschel, 19 March 1848, MM 16, no. 135, and Sabine, *ibid.*, no. 136. As appears from the Council minutes of 13 April it was then decided to request Sabine to approach Rosse, whose letter in reply is Sa, no. 1112. At the meeting of 25 May Sabine was able to report the receipt of letters of acceptance by Rosse to himself and the Treasurer (George Rennie). Peel was in fact approached by de la Beche, but declined promptly and decisively, noting 'That in my opinion the President of the Royal Society ought to be a distinguished man of science; that I thought the departure from the ancient usage had not been successful; that I should oppose any other nomination than a man of science', though de la Beche had said that Herschel and Faraday would decline (Parker, *Sir Robert Peel*, III, p. 492).

39 White's *Journals* for 2, 7, 18, 23 and 30 November 1848; the votes were also reported in *The Athenaeum* for 2 December 1848, p. 1207.

Chapter 4

1 Rosse to Lubbock, Lubbock Correspondence P, no. 85. Rosse's opinion of his business acumen suggested by his query, 'I felt it my duty privately to ask you whether there is any possible risk in his continuing to act as Treasurer to the Royal Society? Whether in the event of his bankruptcy the Society could be a loser, and if so whether there is any step which could be taken to make all safe?' For Rosse's views on the soirées cf. his letters to Sabine in 1848 and 1849, Sa, nos. 1114 and 1115.

2 Council Minutes for 14 June 1855.

3 His letter of 26 October 1854 is not in the Society's archives but can be partially reconstructed from Sabine's reply of 10 November (note 4 below). The same note is struck in his letter of 6 November 1854 to Sabine as Treasurer and Vice-President (MC 5, no. 81; copy in Sa, no. 1123), in his farewell address, printed in *Procs.*, VI, January 1855, and in his earlier letter to Sabine of 23 October 1854 (Sa, no. 1122) when he suggested the names of Argyll, Inglis, Ashburton, Hind, Mantell and Ronalds as Council nominations. A year earlier (26 October 1853; Sa, no. 1120) he had suggested the Earl of Burlington as a Council member with a view to subsequent nomination as President in a letter delightfully revealing of his own genuine commitment to a life of science.

4 Sabine to Rosse, 10 November 1854, President's Letter Book, and 14 November, MC 5, no. 184.

5 See HS 18, nos. 279 and 300. On Wrottesley generally see Layton, 'Lord Wrottesley, F.R.S. Statesman of Science'.

6 White's *Journals* for 30 November 1854.

7 HS 18, no. 301 of 22 July 1847, and Rosse to Sabine, Sa no. 1121, 8 April 1854.

8 The Council Minutes for 25 October 1855 contain both the letter from Wilson, Secretary of the Treasury, and Wrottesley's reply.

9 See White's *Journals* for 20 April 1847, and 12 February, 27 February and 11 March 1852, and Council Minutes for 15 June and 26 November 1852, and below, Chapter 7, as also Council Minutes for 11 October 1866.

10 James South, *A Letter to the Fellows of the Royal and the Astronomical Societies in reply to the Obituary Notice of the Late Rev. Richard Sheepshanks by the President and Council of the Royal Society...with an appendix containing a copy of a memorandum presented by the Earl of Rosse for rendering the Council of the Society more efficient, and which has not been communicated to the Fellows*, 1856. For Private Circulation only. South claimed to have heard of Rosse's memorandum (note 2 above) from his friend A. J. Stephens (F.R.S. 1832), but Stephens was not on the Council. For South's attempts to get a copy of the memorandum, see MC 5, nos. 236, 238 and 250. South also managed to attack his old friend Charles Babbage for his recent book, *The Exposition of 1851*, esp. chs. XII, XIV and XVII, which continued his 1831 diatribes; Babbage was now jealous of South, who had been rewarded by the Government, whereas his own difference engine had never been finished, in spite of much Government money – his own fault.

11 There is an account of the interview in Tyndall, *Faraday as a Discoverer*, pp. 156–58, and a picture depicting the event in the Royal Society's rooms (see frontispiece). See also White's *Journals* for 20 May 1858.

12 See his letter to J. Plücker, 8 April 1856, in L. P. Williams *et al.* (eds.), *Selected Correspondence of Michael Faraday*, p. 834.

13 As noted in his obituary, *Procs.*, 12, for 1862–63.

14 On 13 June 1829, in a leader on the Anatomy Bill.

15 He had proposed retiring a year earlier (MC 6, no. 105) but was persuaded to defer a final resignation for a year, pending an operation. (See Sabine's opinion, MM 19, nos. 22 and 23.) His final letter of 4 June 1861 (MC 6, no. 145) is printed in the Council Minutes for 13 June; it had been read to the meeting of the Society on 6 June.

16 According to the *DNB* it was well known that Graham (who had been Vice-President) had been solicited on this occasion, but that he declined on grounds of ill health.

17 According to White's *Journals* for 19 October 1861, which is confirmed by a letter from Sabine to Sharpey, 30 October 1861, MM 19, no. 32.

18 MC 6, no. 167, letter dated 2 August 1861. (The Council had decided two months earlier.)

19 MM 19, no. 74.

20 So, MC 6, no. 274, Sabine wrote that by charter all matters 'are under the rule of the *President and Council*; not of the President singly'. Huxley was to write to Hooker in 1864 'My distrust of Sabine is as you know chronic...' (Huxley, *Life and Letters of Thomas Henry Huxley*, I, p. 255).

21 Cf. his letters to Sharpey of 21 June (no year), MM 19, no. 51 and 4 July 1861, MC 6, no. 163, both expressing delight in the Council's 'unanimity', and *Procs.* for 30 November 1871.

22 Cf. Huxley, *Life and Letters of Sir Joseph Dalton Hooker*, I, p. 537: 'Though he managed to keep off the very busy Royal Society Council with its heavy work in 1862'.

23 White's *Journals* for 11 January 1868.

24 MM 19, no. 34, proposing G. Everest (F.R.S. 1827) in place of J. Challis (F.R.S. 1848), as not so able, but new to the Council, while Challis had served from 1850 to 1852.

25 MM 19, no. 38, 1 November 1864.

26 White's *Journals* for 31 October 1868. There were Council meetings on 29 October and 5 November to draw up the house-list.

27 MC 8, no. 416, in reply to Granville's letter, which is MC 8, no. 413, dated 26 August 1869.

28 Gassiot, *Remarks on the Resignation of Sir Edward Sabine, K.C.B., of the Presidency of the Royal Society* (Royal Society Tracts 695, no. 8).

29 Gassiot's pamphlet was partly in response to Williamson's action, which he regarded as having been done in an 'offensive manner'. He declared '*An insult offered to a President, particularly at the time when he is occupying the chair*, is *an offence to the Society itself*.' Williamson replied (10 December 1870) in a pamphlet, in which he pointed out that Gassiot was not present and so relied on hearsay, while claiming that he meant no offence to Sabine (Royal Society Tracts X384, no. 5). An anonymous 'contributor' wrote to *Nature* (the lead article for 3 November 1870) describing 'The Government of the Royal Society' to advocate a restriction of the Presidential term to that of ordinary Council members (two years) on the grounds that an indefinite term led to autocracy, to difficulty in change and to the favouring of the President's scientific speciality.

30 White's *Journals* for 4 January 1871. On 19 January Sir James Alderson (F.R.S. 1841, physician) spoke of getting up a petition to Sabine to remain in office for another year while the new officers learned the ropes.

31 White's *Journals* for 9 January 1871. Grove never held office in the Society, to Sharpey's regret according to White (26 June).

32 Thus White noted (16 April 1871) that Dr Haughton of Dublin said that Airy's paper on the perturbations of the earth and Venus 'is one of the grandest achievements of modern science, far beyond the discovery of Neptune. That paper, and his article... in the "Encyclopaedia Metropolitana" place him at the head of science'. He is now regarded as a great scientific organiser rather than a great original scientist.

33 See MC 9, no. 181 for Airy's letter about the cost involved. The Council on 25 May 1871 thanked him 'for the attention he has bestowed on the subject [of the expenses connected with the office of President], and the suggestions which he offers: at the same time as the cost of *Soirées* held in the Society's Apartments will not in future be defrayed by the President, the Council do not anticipate that the expenses attending the Office will be such as to require provision to be made for them by the Society.' See also White's *Journals* for 15, 20 and 24 April and 20 June 1871.

34 W. Airy (ed.), *Autobiography of Sir George Biddell Airy*; p. 293.

35 *Procs.*, 22, 1 December 1873. The

36 See, e.g., MC 9, no. 354 for Crookes's complaint against G. B. Carpenter.

37 See Huxley, *Life and Letters of Sir Joseph Dalton Hooker*, I, pp. 538–44, and McLeod, 'The X-Club a Social Network of Science in Late Victorian England'. The Club broke up in the 1880s when all the members were aging.

38 To Sharpey, 24 October 1856, MM 19, no. 14.

39 Noted by Hooker in the letter to Darwin of 12 January 1873 quoted below, Huxley, *Life and Letters of Sir Joseph Dalton Hooker*, II, p. 133. In 1869 he had told Darwin 'Pray do not C.B your letters to me – I can't stand it. I own C.B gratifies me in a *service* point of view,...but scientifically I rather dislike it' (*ibid.*, p. 147). For his continued reluctance and eventual acceptance see *ibid.*, pp. 147–50.

40 See White's *Journals* for 10 January 1874 for the opinion of 'Mr. Salter' and 23 January 1874 for that of Professor Alfred Newton (F.R.S. 1870). As Lord Burlington the Duke had been a possible candidate some decades earlier.

41 See Council Minutes for 20 March 1873 for Hooker to Spottiswoode, 26 February 1873, 'I assume that, in the event of my election, the Council will not regard the Presidentship as a permanent office. I should wish that the subject of my renomination (if desirable) should be annually carefully considered, with exclusive reference to the interests of the Society; and that at no distant period the Chair should be occupied by a representative of another branch of science.'

42 Huxley, *Life and Letters of Sir Joseph Dalton Hooker*, II, pp. 139–40 (14 January 1875).

43 As in 1877–78, when Dr G. C. Wallich (not F.R.S., although his father had been) accused Huxley, then the junior Secretary, of personal bias (MC 11, nos. 128, 129, 133, 141, 146, 149 (esp.) and 150).

44 Huxley, *Life and Letters of Sir Joseph Dalton Hooker*, II, p. 140. For the address itself, see *Procs.*, 23, 30 November 1874.

45 Council Minutes for 30 November 1875 and *Procs.*, 24, 30 November 1875.

46 To Darwin, 9 June 1878, Huxley, *Life and Letters of Sir Joseph Dalton Hooker*, II, p. 135; also Anniversary address 30 November 1878, *Procs.*, 28, 30 November 1878.

47 See J. H. Gladstone Papers (Royal Society) Hooker's letter of 9 October 1878, tactfully soliciting a donation. The Fund has been variously called Publication Fund, Fee Reduction Fund and Fee Reduction (Publication) Fund. By 1979 its value was about £20,000. The admission fee of £10 was nominally retained but met by the Fund. For details of procedure in 1878/79 see Council Minutes for 19 December 1878, with details of 'composition' which thereafter ceased. For statute amendment see also 16 January 1879.

48 Huxley, *Life and Letters of Thomas Henry Huxley*, I, p. 80.

49 Many Fellows approved (White records the names of Odling, Busk and Savory) while Darwin took the trouble to come to the Society's rooms on 26 February 1872 to support him.

50 White's *Journals* for 25 April 1871: 'Mr. Stokes dreads Huxley's being President, and so accepts Airy'. Stokes's attitude probably lay in the differences in religious outlook of the two men. So White reported for 29 June 1872, 'Dr Beale [F.R.S. 1857, F.R.C.P. 1859] delivered himself of a growl on the prospect of Huxley being Secretary'.

51 Huxley, *Life and Letters of Thomas Henry Huxley*, II, pp. 51–2, 2 July 1883. None of the x Club was on the Council at this time.

52 *Ibid.*, 7 July 1883.

53 *Ibid.*, 14 July 1883.

54 See Council Minutes for 21 May 1885. For his earlier offer to resign, see minutes for 30 October 1884.

55 Huxley, *Life and Letters of Thomas Henry Huxley*, II, p. 175, 1 December 1887, acknowledges his authorship of the leading article in *Nature* for 17 November 1887, 'M.P.P.R.S.'. Several Fellows, including Balfour Stewart (F.R.S. 1862) and A. W. Williamson, supported Stokes

offer to Hooker had been made in the preceding February.

in letters to *Nature* (24 November 1887), while Thistleton-Dyer (F.R.S. 1880, botanist and associate of Hooker) supported the leader on the grounds that Stokes was in an ambiguous position in the House of Commons.

56 Royal Society Tracts X384, no. 4; cf. MC 15, no. 216. Sir Arthur Schuster was later to write (*Biographical Fragments*, pp. 248–57) that 'Wilde would have taken a very prominent place among the scientific men of his time had his exceptional abilities not been handicapped by an obstinate and quarrelsome disposition'.

57 Council Minutes for 16 March, 20 April, 18 May and 15 June 1893. The Committee apparently did not preserve any minutes; the delinquent Fellow was 'remonstrated' with. Cf. also MC 16, no. 16, concerning a dispute over advertisements of Van Houten's cocoa (a new invention) with favourable opinions from J. Attfield (F.R.S. 1880) and S. Ringer M.D. (F.R.S. 1885). Attfield apologised in March 1893 (MC 16, no. 25). For unauthorised use of the designation F.R.S. see Council Minutes for 29 October 1896.

58 The minutes are in CMB 43. The Committee appointed 5 July 1893 included H. E. Armstrong, chemist, F.R.S. 1876; A. W. Rücker, physicist, F.R.S. 1884; E. Ray Lankester, zoologist, F.R.S. 1875; R. T. Glazebrook, physicist, F.R.S. 1882; J. N. Lockyer, astronomer, F.R.S. 1869; W. A. Tilden, chemist, F.R.S. 1880; and the officers. In 1888 Lankester had been much worried about the criteria for publication in *Phil. Trans.*; see MC 14, nos. 320 and 328.

59 There had been a number of proposals to revive the Sectional Committees since their abolition in 1847; fifty years later it seemed safe and advisable to do so.

60 See Council Minutes for 16 May 1895, when the possibility of accepting paid advertising was discussed, and for 15 and 20 June and 5 July 1895, about applying to the Treasury for a publication grant.

61 Quoted in Strutt, *John William Strutt*,

p. 173, where Rayleigh's letter to *The Times* is also printed.

62 Thompson, *The Life of William Thomson Baron Kelvin of Largs*, p. 945.

63 Fisher, *Joseph Lister*, p. 298.

64 *Ibid.*, p. 299.

65 See Lydekker, *Sir William Flower*. The author believed that Flower had refused to be nominated because he was too busy and it seems likely that it was on this occasion. Flower had received the Royal Medal in 1882.

66 Leader on 7 December 1895, pp. 1443–44.

67 He was not the only possible candidate. Besides Graham there were others: Col. Philip Yorke who had been president of the Chemical Society was mentioned, while Gassiot proposed Faraday to which Sabine replied, 'I *could* not ask Faraday to take a place which in some sort might seem second to myself...I greatly doubt that any solicitation would prevail with Faraday: if it would, well: – otherwise I should be right well pleased with either Yorke or Miller' (Sabine to Sharpey, 30 October 1861, MM 19, no. 32).

68 For a favourable view, see Taylor, 'The Life and Teaching of William Sharpey (1802–1880)'. Sharpey was never popular with *The Lancet*, because Wakley thought that Grant should have had the Chair, but he was not openly attacked there until he was much blamed for the fracas over Lee's grievances in 1855.

69 Foster to Rayleigh, in Strutt, *John William Strutt*, p. 168.

70 Sabine to Sharpey, 26 March 1861, MM 19, no. 25.

71 The correspondence is in MC 5, nos. 84, 89, 90, 97, 100. See also *Official Report of the Fall of a Thunderbolt or Meteorolite* [sic] *at Dover...with a series of lectures on the nebulous origin...of the Earth...with...the causes of magnetic attraction* [by Higginson], London, 1853. Christie, confirmed by Waterhouse of the British Museum, found that the stones were not meteoric, but iron pyrites commonly found in chalk and evidently washed up on Dover Beach.

72 MC 9, no. 104.

73 MC 10, nos. 71, 79, 83; the paper was on the dimensions of the Great Pyramid. Cf. the complaints by a Fellow because he thought that Stokes and others were partisan in a controversy with Carpenter, as retailed by White, 23 and 25 October 1877: 'I asked Mr. Stokes if he had written to Carpenter that he thought C. [*sic*] wrong. "No. All I wrote was, that C. put forward no theory, but seemed to lean to the idea that the movements were due to the direct action of light."'

74 See Council Minutes for 5 July and 30 November 1895, MC 16, nos. 227, 255, 269, the last, to Rayleigh, the junior Secretary, concerned with Pavy's sense of injustice because his refutation of Paton had been archived, only an abstract having even been read. Foster's letter was printed in *The Lancet* for 28 December 1895 (pp. 1664–65) entitled 'The Publication of Papers read before the Royal Society', Pavy's reply appearing in the issue for 4 January 1896; in a judicious review in the issue for 29 February 1896 *The Lancet* editorially regarded the matter as settled. Paton's paper had been refereed during the summer; see Referees' Reports RR 12, nos. 203 and 204 (reports by J. W. McKendrick and W. D. Halliburton).

75 MC 12, no. 167, to the Assistant Secretary.

76 Council Minutes for 2 June 1853. It had been proposed (30 May and 13 June 1852) that two medals were one too many, but this was voted against. For comments by Airy, H. Lloyd, Baden Powell, Herschel and R. Brown see MC 4, nos. 334, 335, 338, 340 and 342.

77 Letter to Christie, 15 November 1853, MC 5, no. 123, printed in the Council Minutes for 24 November 1853, and to Gassiot, MC 5, no. 124. Cf. William Hopkins (F.R.S. 1837) writing from Cambridge to approve Tyndall's action, MC 5, no. 125; Hopkins was on the Council but not present when the award was voted.

78 In 1902 new statutes abolished the right of privy councillors but added a provision for the election of persons who 'either have rendered conspicuous service to the cause of science or are such that their election would be of signal benefit to the Society'.

Chapter 5

1 For modern regulations, see the *Yearbook*; for the changes made in 1831, see *Record*, 4th edn, p. 113.

2 It is printed in Council Minutes, II, pp. 138–39.

3 It is interesting to compare the modern regulations (*Yearbook*, 1979, Council Procedure): 'The grant for scientific investigation is provided to promote and support research in science and to assist scientific expeditions and collections; it is not intended for personal maintenance, payment of stipends or to aid scientific publications.'

4 Todhunter, *William Whewell*, II, p. 357, letter of 17 March 1850, showing that the committee was to hold its first meeting that week. For later changes in membership see Council Minutes for 27 January 1859. The British Association had begun giving small grants in 1834 (for studies on the tides); the monies came from the subscription fees of those attending meetings, about half of which were eventually applied to research. But it cannot have distributed anything like £1000 p.a. before about 1870.

5 Lubbock Correspondence P, no. 87 (copy).

6 Second leader 8 December 1849; presumably Wakley also disapproved of the medical men, who included John Forbes and Roget; besides those already named were J. J. Bennett, botanist, and Henry Moseley, mathematician. All the Council were eminent enough to have entries in the *DNB*.

7 Report to the Council on 17 February 1876.

8 Council Minutes for 18 May 1876 with a report of a letter from the Lord President of the Council offering 'further aid to be given to research by according permission to the Government Grant Committee to recommend in certain cases the payment of personal allowances to

gentlemen during the time they are engaged in their investigations' (cf. *Record*, pp. 186–87). The letter also made certain proposals about administration and accountability. All the scientists who gave evidence on the matter to the Devonshire Commission had declared their belief that the Grant should be increased, with the single exception of Airy, pessimistically sure that to ask for more might cause the Government to refuse to give anything (cf. *Nature*, 12, 1875, 391).

9 Council Minutes for 14 December 1876.

10 The Royal Society's own list contained the names 'of those Fellows of the Royal Society who have at any time received parcels of books consigned for them to the Royal Society, all who have written papers printed in the Philosophical Transactions, all who have been on the Council and on Scientific Committees', as is noted in the Council Minutes for 22 April 1852. See also MC 5, esp. no. 47.

11 Preface to Volume I (London, 1867), which contains a brief account of the pre-history of the project.

12 Besides the accounts in the *Catalogue* itself (note 11) there is one in the *Record*, pp. 180–82; and, very brief, in Lyons, p. 185. The origins are revealed in Council Minutes, esp. for 5 March 1857, and the Report of the Library Committee, 14 January 1858 (but there is no reference to any pre-history there), as well as the Presidential address in 1858. Because of the conventions of the time (first names not given), it is impossible to be sure which Grant (I have assumed Robert) and which De Morgan (Augustus, mathematician, probably) were in fact on the Committee.

13 The committee's minutes are in CMB 1.

14 The minutes cover 1838 and 1839; from 1839 to 1845 the committee was called 'Physics with Meteorology'. Rough minutes, drafts and related correspondences may be found in DM 3, nos. 2–118. Also see, for 1856, CMB 42.

15 Airy to P.R.S., 28 June 1840, MM 11,

no. 145; the editor of *The Athenaeum* had been eager to print them in 1834.

16 The letter, which is AP 20, no. 7 (in French, translation no. 8), was 'laid before' the Council on 5 May 1836, when Airy and Christie were requested to consider it and report, which they did (favourably) on 9 June 1836. The report was then read to the Society, printed in *Procs.* and 250 extra copies run off for distribution. The subsequent committee to consider 'the best means of carrying into effect the measures recommended' consisted of the Treasurer (Baily), the Secretaries (Roget and Children), Christie, Lubbock and Whewell, but after the appointment of Sectional Committees, the Joint Committee of Physics and Meteorology (of which Herschel was chairman) took over responsibility for the matter. See also the report of the Joint Committee of Physics and Meteorology printed in the Council Minutes for 22 December 1838, and below, Chapter 8. The Committee insisted that 'horizontal direction, the dip and intensity, require to be precisely ascertained before the magnetic state of any given station on the globe can be said to be fully determined'.

17 As by Nathan Reingold, article on Sabine in the *DSB*, who also claims that it was because the British Association was unable to influence the Government that Sabine persuaded Humboldt to write to the Royal Society. Cf. Cawood, 'The Magnetic Crusade: Science and Politics in Early Victorian Britain'.

18 See *Record*, esp. pp. 198–99, and Howarth, *The British Association*, pp. 155–69.

19 See Council Minutes for 15 June 1854; the Society's response was approved by the Council on 22 February 1855; some of the minutes for the Met. Committee after 1856 are in CMB 42. The whole early history is reviewed in the 1866 printed Report (see below p. 158). See also Mellersh, *Fitzroy of the Beagle*, pp. 262–87.

20 Correspondence about setting up the new Met. Committee is printed with the Council Minutes for 13 December 1866. For its ending see Council

Minutes for 12 April 1877, 7 and 14 June 1877 and 25 October 1877. See also Galton, *Memories of my Life*, ch. XVI; copies of all the published reports and of the relevant sections of the Devonshire Commission report are preserved in the Galton MSS. 118/1 in the library of University College London.

Chapter 6

1 Properly 'The Science Commission Report on the Advancement of Science'; the Commission was under the chairmanship of the 7th Duke of Devonshire who as the Earl of Burlington had more than once been considered as a possible P.R.S.

2 Both his original letter (read at the Council meeting of 16 November 1849) and Rosse's reply have disappeared. Rosse regarded both as confidential and did not have a copy made for the Society's archives and by the summer of 1855 claimed to have lost both (see his letter MC 5, no. 206). Part is quoted in the Council Minutes and part in the report of the Committee (also in the minutes). Both were partly quoted by Sabine in his evidence before the Devonshire Commission and given *in extenso* in his written report, printed as vol. II, appendix XI (pp. 41–62), whence it was quoted in *Nature*, 12, 1875, 285–88, 305–8, 360–64, 389–92, 469–70. Cf. McLeod, 'The Royal Society and the Government Grant: Notes on the Administration of Scientific Research 1849–1941', useful but unreliable in its references and making no *direct* use of the Devonshire Commission report.

3 Some members of the Government tried to accuse the Society of assuming as an annual right what had been intended as only a unique gift, and even to imply that the initiative had come from the Society in the first place, but this was readily repudiated.

4 The letter of the Secretary of the Treasury and Wrottesley's reply are printed in the Council Minutes for 25 October 1855.

5 Cf. *The Lancet* for 4 August 1855, p. 116, 'Professor Owen and the Royal Society'.

6 Cf. letters in MC 8, nos. 346 and 350 of April 1859 from Fellows anxious to have the President and Council emphasise this point to the public.

7 This of course continued.

8 Cf. Council Minutes for 27 October 1853.

9 Council Minutes for 21 February 1867.

10 It consisted of the Astronomer Royal, Rosse, Wrottesley, Sir John Shaw Lefevre, Sabine, Graham and W. H. Miller. Perhaps oddly in view of its interest in standardisation, the Royal Society does not seem to have been associated with the British Association's work on the standardisation of screw threads (1881–84).

11 For the committee's report, see *Procs.*, 7, 1854–55, 499–509. Willis was Jacksonian Professor of Applied Mechanics at Cambridge. Babbage looked favourably upon Schütz and his machine.

12 The Council Minutes for 18 May 1865 contain all letters and memoranda at length. See, for details, Cotter, 'The Royal Society and the Deviation of the Compass', and Council Minutes for 2 November and 30 November 1865.

13 Council Minutes for 18 January 1866.

14 Council Minutes for 18 February, 18 March, 27 October and 24 November 1869.

15 Council Minutes for 18 February 1869.

16 Council Minutes for 27 February 1879.

17 Council Minutes for 25 April 1895.

18 Council Minutes for 25 October 1888.

19 Council Minutes for 19 March 1896; the original letter of enquiry is MC 16, no. 299.

20 Council Minutes for 18 July 1876. Cf. French, *Antivivisection and Medical Science in Victorian Society*.

21 *Procs.*, 30 November 1880. The problem concerned both the *method* of vaccination and the question of compulsory vaccination. No new Bill was passed to replace that of 1853, in spite of a long inquiry, for over ten years.

22 Council Minutes for 28 April 1892 for the final report.

23 Council Minutes for 28 May 1891; the

minutes of the Committee are in MS. 511.

24 Council Minutes for 2 July 1896 and CMB 5 (minutes).

25 Council Minutes for 28 January 1897.

26 Council Minutes for 7 July 1898 and CMB 14, CMB 15.

27 Council Minutes for 13 May and 15 June 1852.

28 Council Minutes for 18 June 1868.

29 Council Minutes for 15 March 1883. At Huxley's urging it was decided to apply to the War Office for assistance and to offer £150 from the Donation Fund. By 5 July the Council had received a letter from the War Office enclosing 'a detailed report...by Colonel Heriot Maitland, Commanding Royal Engineers in Egypt'. The Council approved his report and asked the War Office to 'instruct' him to proceed with all speed before the seasonal rising of the Nile. He was to be assisted by Major R. H. Williams, and, with a further £200 from the Donation Fund, work continued for four years, although hampered by a cholera epidemic in 1884. The final report (Council Minutes for 27 October 1887; see also the minutes of the Delta Committee in CMB 3) described how the boring had gone down to 308 feet without striking rock, but providing much information about the stratigraphy. Although the Committee would have liked the work to continue it does not seem that any more was done.

30 Council Minutes for 17 October 1895; the Coral Reef Committee minutes are in CMB 8. The whole enterprise began with an application for a government grant from W. J. Sollas (F.R.S. 1889), interested in the problem from a geological point of view; the Government Grant Committee in accepting the application and offering £800 (16 May 1895) recommended that application should be made by the President and Council to the Admiralty for a surveying vessel. The final result was published as a monograph by the Coral Reef Committee (chairman T. G. Bonney, F.R.S. 1878, professor of geology at University College London).

31 See especially p. 2. The author quotes Babbage, the letters of W. R. Hamilton (F.R.S. 1813, an antiquary) to Lord Elgin, and James Millingen (archaeologist).

32 Its proceedings and final report were published *in extenso* as Parliamentary Papers, the first in 1870, the final (eighth) report in 1875.

33 Wrottesley had reported on this at the Anniversary Meeting in 1856 (*Procs.*, **VIII**, 1856–57). See Council Minutes for 15 January 1857.

34 In the Eighth Report there is a list of 'scientific work carried on by Departments of Government'. Only eight examples are given: Topographical Survey (Treasury), Hydrographical Survey (Admiralty), Geological Survey (Privy Council), Astronomical Observations (Admiralty and Treasury), Meteorology (Greenwich Observatory, Treasury (Edinburgh), Met. Office), Botany (gardens), Chemistry (Department of War Office), Standards (Department of Board of Trade). It was stated that the existing organisation was 'insufficient', and hence, in part, the call on the Royal Society for advice and assistance. (No mention is made of the Patent Office, which the Royal Society was nearly called upon to overhaul in 1868; see MC 8, nos. 169 and 179.)

35 Eighth Report, pp. 45–46.

36 *Procs.*, **60**, also *Yearbook* 1896–97.

Chapter 7

1 For the presidents (with brief biographies) see Howarth, *The British Association*, appendix II, pp. 293–303; the tradition continued in the twentieth century. Morrell and Thackray in *Gentlemen of Science* see it differently, but supply similar data. John Herschel's reputation was high in both institutions, and he served as president of the British Association in 1845 at the second Cambridge meeting, but he was not particularly more active otherwise than as F.R.S.

2 *Procs.* III (1830–37), no. 11, pp. 140–45.

3 See Howarth, *British Association*, appendix I, 'Classified List of Grants made by the British Association in Aid

of Research', p. 217; the grant of £247 12s 5d was for 'Instruments etc'. Morrell and Thackray, *Gentlemen of Science*, pp. 313–24, discuss the distribution of grants, revealing that in general only a few men (F.R.S.) received the majority of them.

4 It was proposed by T. R. Robinson (F.R.S. 1856), director of Armagh Observatory, president of the British Association, in a letter to Rosse as P.R.S. in October 1849, and approved at a Council meeting on 11 April 1850; the British Association then sent a copy of their 'Memorial' to Lord John Russell to the Society (Council Minutes for 19 December 1850) with a letter signed by the secretaries, one of whom was Sabine, by now Treasurer of the Royal Society.

5 Council Minutes for 11 November 1841, 10 February 1842 and 10 March 1842.

6 See esp. Howarth, *The British Association*, pp. 155–68.

7 See, *inter alia*, Council Minutes for 11 July 1856, 11 October 1866, 15 December 1870, 5 November 1896. The Gassiot Fund now goes to the Meteorological Office.

8 The Council Minutes for 1 May 1884 contain the letter of the Astronomer Royal 'explaining the position of England in regard to the Metrical Convention'. The Council accepted his advice and sent a delegation to the Treasury which, after difficulties, agreed to the proposal (Council Minutes for 29 May, 3 July and 18 December 1884, and *Procs.*, 37, 1884 Treasurer's address at the Anniversary Meeting, which contains, p. 434, a summary). In 1887 the Treasury decided to withdraw but the Council, alerted again by the Astronomer Royal, urged again 'the Scientific interests of the country' and were successful (Council Minutes for 7 July and 27 October 1887).

9 Cf. Presidential address for 1889, *Procs.*, 46, 1889.

10 As appears from a letter written by Lister (MC 15, no. 186) part of the difficulty was to ensure that it was not taken to be a money-making institution, as it might be if the title included the word 'Limited' as the

Board of Trade at first wished, but pressure by anti-vivisectionists was also very strong (see Godlee, *Lord Lister*, pp. 495–97).

11 Bonney, *Annals of the Philosophical Club*, s.v. 27 November 1847, 27 January 1848 (when a committee reported), 22 May 1848, 24 January, 28 February and 28 March 1850, 1852 and 1853 *passim*.

12 Council Minutes, esp. for 15 June and 22 June 1852, 26 November 1852, 20 January 1853, and Philosophical Club minutes. Sabine, Treasurer, was on the Philosophical Club's committee, as was Bell; Daubeny, president of the Chemical Society was on the Council of the Royal Society; and so on. It was all a trifle Gilbertian.

13 See Council Minutes for 20 January 1853 printing the Astronomical Society's reaction to the Second Report (p. 34) of the 1851 Commissioners.

14 Philosophical Club minutes for 30 April 1855, Bonney, *Annals of the Philosophical Club*, pp. 43–44.

15 For the final decision, and moves leading to it, see MM 13, no. 59 and Council Minutes for 30 May 1856; nos. 60–78 are minutes of the Burlington House and Removal Committees.

16 See the newspaper report of the debate in MM 13, no. 64.

17 See Council Minutes, esp. for May 1858. There were a dozen societies which sought accommodation, about half of them scientific.

18 On all these points see Council Minutes, esp. for 11 October 1866, 3 June and 2 July 1868; White, *Journals*; also MC 16, no. 112.

19 See Huggins, *The Royal Society*, 1903 address.

20 Council Minutes for 6 November 1890 and 19 February 1891. See also MM 16, nos. 28 and 30; the latter is a 'Draft of the proposed' Bill.

21 Council Minutes for 6 November 1879 and 16 June and 21 October 1881 and MC 18, no. 237. When the Royal Society wished to sell its Acton lands in 1880 the case had to be argued before the courts, but the Society successfully showed that the land had been purchased from income and was

not part of the endowment, so that the sale was legal even for a charity.

22 Cf. Donnan, 'The Scientific Relief Fund and its Committee'.

23 Todhunter, *William Whewell*, II, p. 357.

24 *Yearbook* for 1899, p. 135 and for 1900, p. 137 give the history in outline. See also MC 17, no. 55, August 1897, where Schuster reported on the meeting of the 'Kartell'. The first delegates were Schuster (German by birth and education, but patriotically English) and H. E. Armstrong (who had taken a German Ph.D. in chemistry). See also Council Minutes for 8 April 1897, 28 October 1897, 5 May 1898 and 16 June 1898. Cf., especially for later activities, Alter, 'The Royal Society and the International Association of Academies, 1897–1919'.

25 Council Minutes for 11 July 1856. The Cambridge Commissioners were Peacock, Herschel, Sedgwick and the Attorney-General.

26 Its report was presented on 15 January 1857.

27 Council Minutes for 16 January and 20 February 1890.

28 Council Minutes for 4 March 1897.

29 See his 1905 Presidential address and the appendix on 'Science in Education' in his *The Royal Society*.

Chapter 8

1 The Council Minutes provide information about correspondence (sometimes printed in full) between the Society, the Admiralty, and the individuals involved. For secondary sources I have found useful McConnell, *Historical Instruments in Oceanography* and *No Sea Too Deep*; Thomson, *The North-West Passage*; Dodge, *The Polar Rosses*; and biographies of the ships' captains in the *DNB*.

2 The Pendulum Committee had been involved in the work by Kater and others to determine the length of a second's pendulum, partly for determining standards of length, partly to determine the exact shape of the earth. Its minutes are in CMB 1 and DM 3.

3 John Ross's *A Voyage of Discovery* and *Narrative of a Second Voyage in Search of a North-West Passage* contain many scientific reports, including (as the second) J. C. Ross's discovery of the North Magnetic Pole, as well as accounts of natural history and the language of the Eskimos. For the resultant controversy see Sabine, *Remarks on the Account of the late Voyage of Discovery to Baffin's Bay published by Captain J. Ross, R.N.* and John Ross, *An Explanation of Captain Sabine's Remarks on the Late Voyage of Discovery to Baffin's Bay*. The Pendulum Committee at its meeting of 18 March 1819 had congratulated Sabine on his work; his accounts of magnetism, ornithology and anthropology were to be published in *Phil. Trans.*

4 Council Minutes for 15 June, 17 June and 16 November 1826. This was Parry's last voyage for he had been appointed hydrographer to the navy in 1823.

5 He sent a paper about this to the Royal Society, which is published in *Procs.*, 1834, as well as in his uncle's account.

6 Council Minutes and minutes of the Physics and Meteorology Committee. There are two chief primary accounts: that by James Clark Ross, *A Voyage of Discovery and Research in the Southern and Antarctic Regions*, and Joseph Hooker's letters in Huxley's *Life and Letters of Sir Joseph Dalton Hooker*, chs. II–VII (pp. 37–167). Hooker produced a summary for the official account of some of the scientific results: Richardson and Gray (eds.), *The Zoology of the Voyage of HMS Erebus & Terror…1839 to 1843*.

7 According to Friendly, *Beaufort of the Admiralty*, Fitzroy asked Beaufort to recommend a young naturalist; Beaufort in turn approached Peacock, who asked Henslow.

8 See Day, *The Admiralty Hydrographic Service, passim*.

9 For the correspondence of Sabine and Herschel, see HS 2, nos. 24 and 25 (23 November and 3 December). Friendly (note 7 above) argues that Barrow was the major obstacle, because he wanted all voyages to go to the Arctic; he

notes that a new Government would
have meant the appointment of a new
First Lord of the Admiralty who
might or might not be sympathetic. If
Peel had been the Prime Minister
concerned he might well have been
interested in the voyage, but clearly
the political excitements of a change of
Government would not have been
favourable. Cannon, 'History in
Depth: the Early Victorian Period',
insists that the Bed Chamber Crisis
and Peel's consequent failure to form a
Government is involved here, but the
whole matter was settled before
Melbourne's Government fell.

10 See the minutes of the Committee for
Botany and Vegetable Physiology, and
also Huxley, *Life and Letters of Sir
Joseph Dalton Hooker*, I, p. 44.

11 See Wilson and Geikie, *Memoir of
Edward Forbes F.R.S.*

12 See Deacon, *Scientists and the Sea
1650–1900*: after reviewing the
oceanographic aspects of earlier
voyages she devotes ch. 14 to
Thomson, Carpenter and the
Challenger expedition; and Schlee, *A
History of Oceanography*, ch. II, pp.
107–38.

13 See Council Minutes for 13 June 1844
and 3 April 1845.

Chapter 9

1 See Lyons, appendix III, p. 343.

2 Thus the young Barclay Fox in 1840
reported 'We strangers entered as our
names were called and were
entertained by the dull proceedings of
this dull society' (which included a
paper by Daniell on electrolysis,
although Fox does not say so). But
although an acquaintance of Davies
Gilbert, and having attended meetings
of the British Association, Fox was
clearly not interested in science, to
which he seems to have been exposed
because his sister was interested in it
and dragged him into scientific
society. Cf. Brett (ed.), *Barclay Fox's
Journal*, esp. 2 June 1838 and 21 May
1840, with Pym (ed.), *Memories of Old
Friends*, extracts from the journal and
diaries of Caroline Fox, esp. 1836 and
2 and 5 June 1838. She listened with
interest to Wheatstone's conversation
at King's, and clearly would have
enjoyed the Royal Society meeting
which so bored her brother.

3 The 1981 Annual Report (of the
Council) began by quoting from the
retiring President's address. The other
'objects' cited by Todd were 'to offer
the Government an independent
source of advice and help' and 'to
uphold and develop international
scientific relations', all themes familiar
to his predecessors of the later
nineteenth century.

BIBLIOGRAPHY

Airy, W. (ed.), *The Autobiography of Sir George Biddell Airy* (Cambridge, 1896)

Alter, Peter, 'The Royal Society and the International Association of Academies, 1897–1919', *Notes and Records of the Royal Society*, **34**, 1980, 241–64

Anon, *Defence of the Resolution for Omitting Mr. Panizzi's Bibliographical Notes from the Catalogue of the Royal Society* (1838) [probably by Rigaud]

– Review of Barrow's *Sir Joseph Banks and the Royal Society* in Sharpe's *London Magazine*, 1845 (221–35, 351–69, 565–86)

– *The Record of the Royal Society of London*, 4th edn (London, 1940)

Babbage, Charles, *The Exposition of 1851* (London, 1851, reprinted 1968)

– *Passages from the Life of a Philosopher* (London, 1864)

– *Reflections on the Decline of Science in England* (London, 1830)

Baily, Francis, *Remarks on the Present Defective State of the Nautical Almanac* (London, 1822)

– *Astronomical Tables and Remarks for the Year 1822* (London, 1822)

– *Further Remarks on the Present Defective State of the Nautical Almanac for 1822* (London, 1829)

Barrow, Sir John, *Sketches of the Royal Society and Royal Society Club* (London, 1849)

Bonney, T. G., *Annals of the Philosophical Club of the Royal Society* (London, 1919)

Brayley, E. W., *The Utility of the Knowledge of Nature considered; with reference to the introduction of instruction in the Physical Sciences into the General Education of Youth* (1831)

Brett, R. L. (ed.), *Barclay Fox's Journal* (London, 1979)

Cameron, H. C., *Sir Joseph Banks* (London, 1952)

Cannon, Walter F., 'History in Depth: the Early Victorian Period', *History of Science*, **3**, 1964, 20–38

Cawood, John, 'The Magnetic Crusade: Science and Politics in Early Victorian Britain', *ISIS*, **70**, 1979, 493–518

Cooper, M. L. and Hall, V. M. D., 'William Robert Grove and the London Institution, 1841–1845', *Annals of Science*, **39**, 1982, 229–54

Cotter, Charles H., 'The Royal Society and the Deviation of the Compass', *Notes and Records of the Royal Society*, **31**, 1977, 297–310

Curwen, E. C., *The Journal of Gideon Mantell* (London, 1940)

Daniell, J. F., *An Introductory Lecture* [delivered at King's College, London] (1831)

Day, Archibald, *The Admiralty Hydrographic Service 1795–1919* (London, 1967)

Deacon, Margaret, *Scientists and the Sea 1650–1900* (London, 1971)

Devonshire Report: *see* Parliamentary Papers

Dodge, E. S., *The Polar Rosses. John and James Clark Ross and their Explorations* (London, 1973)

Donnan, F. G., 'The Scientific Relief Fund and its Committee', *Notes and Records of the Royal Society*, 7, 1950, 158–71

Dreyer, J. L. E. and Turner, H. H. (eds.), *The History of the Royal Astronomical Society 1820–1920* (London, 1923)

Fisher, R. B., *Joseph Lister 1827–1912* (London, 1977)

[Fitton, W. H.], *A Statement of Circumstances Connected with the late Election for the Presidency* (1831)

French, Richard D., *Antivivisection and Medical Science in Victorian Society* (Princeton, 1975)

Friendly, Alfred, *Beaufort of the Admiralty. The Life of Sir Francis Beaufort, 1774–1857* (London, 1977)

F.R.S., *Thoughts on the Degradation of Science in England* (1847)

Galton, Francis, *Memories of my Life* (London, 1908)

Gassiot, J. P., *Remarks on the Resignation of Sir Edward Sabine K.C.B. of the Presidency of the Royal Society* (1870)

Geikie, Sir Archibald, *The Annals of the Royal Society Club* (London, 1917)

Gilbert, L. F., 'The Election to the Presidency of the Royal Society in 1820', *Notes and Records of the Royal Society*, 11, 1954, 256–79

Gillen, Mollie, *Royal Duke. Augustus Frederick Duke of Sussex* (London, 1976)

Godlee, R. J., *Lord Lister* (London, 1917)

Granville, Augustus Bozzi, *The Royal Society in the XIXth Century* (London, 1836). A second edition of *Science without a Head, or the Royal Society Dissected by One of the 687 F.R.S. ---sss* (London, 1830)

Granville, Paulina B. (ed.), *Autobiography of A. B. Granville* (London, 1874)

Hall, A. Rupert and M. B. (eds.), *The Correspondence of Henry Oldenburg*, vols. I–XIII (Madison, Milwaukee and London, 1965–84)

Hall, Marshall, *A Letter addressed to the Earl of Rosse, President-elect of the Royal Society*, 2nd edn (1848)

Howarth, O. J. R., *The British Association for the Advancement of Science: A Retrospect 1831–1936*, 2nd edn (London, 1931)

Huggins, Sir William, *The Royal Society: or Science in the State and in the Schools* (London, 1906)

Huxley, Leonard, *Life and Letters of Sir Joseph Dalton Hooker*, 2 vols. (London, 1918)

– *Life and Letters of Thomas Henry Huxley*, 2 vols. (London, 1900)

Jones, Bence, *The Life and Letters of Faraday*, 2nd edn, 2 vols. (London, 1870)

Larmor, Sir Joseph, *Memoir and Scientific Correspondence of George Gabriel Stokes* (Cambridge, 1907)

Layton, David, 'Lord Wrottesley, F.R.S. Statesman of Science', *Notes and Records of the Royal Society*, 23, 230–46

Lee, Robert, *Memoirs on the Ganglia and Nerves of the Uterus* (London, 1849)

Lover of Science, *The Knighthood of Sir J. South A Death Blow to Science* (1830)

Lydekker, R., *Sir William Flower* (London, 1906)

Lyell, K. M. (ed.), *Memoir of Leonard Horner*, 2 vols. (London, 1890)

Lyons, Sir Henry, *The Royal Society, 1660–1940* (Cambridge, 1944)

McConnell, Anita, *Historical Instruments in Oceanography* (London, 1981)

– *No Sea Too Deep. The History of Oceanographic Instruments* (Bristol, 1982)

McLeod, R. M., 'Of Medals and Men: A Reward System in Victorian Science 1826–1914', *Notes and Records of the Royal Society*, **26**, 1971, 81–105

– 'The X-Club a Social Network of Science in Late Victorian England', *Notes and Records of the Royal Society*, **24**, 1970, 305–12

– 'The Royal Society and the Government Grant: Notes on the Administration of Scientific Research 1849–1941', *Historical Journal*, **14**, 1971, 323–55

Mellersh, H. E. L., *Fitzroy of the Beagle* (London, 1968)

Miller, D. P., 'Between Hostile Camps: Sir Humphry Davy's Presidency of the Royal Society of London, 1824–27', *British Journal for the History of Science*, **16**, 1983, 1–47

Moll, Gerrit, *On the Alleged Decline of Science in England* (London, 1831)

Morrell, Jack and Thackray, Arnold, *Gentlemen of Science. Early Years of the British Association for the Advancement of Science* (Oxford, 1981)

Panizzi, Anthony, *A Letter to His Royal Highness the President of the Royal Society on the New Catalogue of that Institution now in Press* [1837/38]

– *Observations on the Address by the President and on the Statement by the Council...* (1837)

Paris, J. A., *The Life of Humphry Davy* (London, 1831)

Parker, Charles Stuart, *Sir Robert Peel*, 3 vols. (London, 1899)

Parliamentary Papers: the Royal Commission on Scientific Instruction and the Advancement of Science. [Or] The Scientific Commission *Report on the Advancement of Science* [the Devonshire Commission Report], 8 vols. (1872–75), 2213–15

Peacock, George, *The Life of Thomas Young* (London, 1855)

Pym, H. N. (ed.), *Memoirs of Old Friends* (London, 1882)

Ramsden, Lady Guendolen, *The Correspondence of Two Brothers* (London, 1906)

Reingold, Nathan, 'Babbage and Moll on the State of Science in Great Britain', *British Journal for the History of Science*, **4**, 1968, 58–61

Richardson, Sir John and John Edward Gray (eds.), *The Zoology of the Voyage of HMS Erebus & Terror...1839 to 1843* (London, 1846)

Ross, James Clark, *A Voyage of Discovery and Research in the Southern and Antarctic Regions...*, 2 vols. (London, 1847)

Ross, John, *Narrative of a Second Voyage in Search of a North-West Passage* (London, 1835)

– *A Voyage of Discovery...* (London, 1819)

– *An Explanation of Captain Sabine's Remarks on the Late Voyage of Discovery to Baffin's Bay* (London, 1819)

Rudwick, Martin, 'The Foundation of the Geological Society of London', *British Journal for the History of Science*, **1**, 1963, 325–55

Sabine, Edward, *Remarks on the Account of the late Voyage of Discovery to Baffin's Bay published by Captain J. Ross, R.N.* (London, 1819)

Schlee, Susan, *A History of Oceanography* (London, 1975)

Schuster, Arthur, *Biographical Fragments* (London, 1932)

South, Sir James, *Charges against the President and Council of the Royal Society* (1830)

– *Refutation of the numerous mis-statements and fallacies in a paper presented to the Admiralty by Dr. Thomas Young* (1829)

– *A Letter to the Fellows of the Royal and the Astronomical Societies in reply to the Obituary Notice of the late Rev. Richard Sheepshanks*...(1856)

Stimson, Dorothy, *Scientists and Amateurs – A History of the Royal Society* (New York, 1948)

Strutt, R. J., *John William Strutt, 3rd Baron Rayleigh, O.M., F.R.S.* (London, 1924)

Taylor, D. W., 'The Life and Teaching of William Sharpey (1802–1880)', *Medical History*, 15, 1971, 421–59

Thompson, S. P., *Michael Faraday his Life and Work* (London, 1901)

– *The Life of William Thomson Baron Kelvin of Largs*, 2 vols. (London, 1910)

Thomson, G. M., *The North-West Passage* (London, 1875)

Todd, A. C., *Beyond the Blaze. A Biography of Davies Gilbert* (Truro, 1967)

Todhunter, I., *William Whewell*, 2 vols. (London, 1876, facsimile reprint 1970)

Tyndall, John, *Faraday as a Discoverer* (London, 1868)

Weld, C. R., *A History of the Royal Society*, 2 vols. (London, 1848)

[W. White (ed.)], *The Journals of Walter White* (London, 1898)

Williams, L. P., *Michael Faraday* (London, 1965)

– 'The Royal Society and the Founding of the British Association for the Advancement of Science', *Notes and Records of the Royal Society*, 16, 1961, 221–33

Williams, L. P., Fitzgerald, R. and Stallybrass, O. (eds.), *The Selected Correspondence of Michael Faraday*, 2 vols. (Cambridge, 1971)

Williamson, A. W., *Mr. Gassiot and the Royal Society* (London, [the author], 1870)

Wilson, George and Geikie, A., *Memoir of Edward Forbes, F.R.S.*....(Cambridge, London, Edinburgh, 1861)

Wood, A., *Thomas Young Natural Philosopher* (Cambridge, 1954)

INDEX

Printed in the United States

Printed in the United States
By Bookmasters